BIOTECHNOLOGY
INTELLIGENCE
UNIT

PLANT BIOTECHNOLOGY TRANSFER TO DEVELOPING COUNTRIES

David W. Altman

UST
Nashville, Tennessee, U.S.A.
Cornell University
Ithaca, New York, U.S.A.

Kazuo N. Watanabe

Cornell University
Ithaca, New York, U.S.A.

R.G. LANDES COMPANY
AUSTIN

BIOTECHNOLOGY INTELLIGENCE UNIT

PLANT BIOTECHNOLOGY TRANSFER TO DEVELOPING COUNTRIES

R.G. LANDES COMPANY
Austin, Texas, U.S.A.

Please address all inquiries to the Publisher:
R.G. Landes Company, 909 Pine Street, Georgetown, Texas, U.S.A. 78626
or
P.O. Box 4858, Austin, Texas, U.S.A. 78765
Phone: 512/ 863 7762; FAX: 512/ 863 0081

U.S. and Canada ISBN 1-57059-238-1

While the authors, editors and publisher believe that drug selection and dosage and the specifications and usage of equipment and devices, as set forth in this book, are in accord with current recommendations and practice at the time of publication, they make no warranty, expressed or implied, with respect to material described in this book. In view of the ongoing research, equipment development, changes in governmental regulations and the rapid accumulation of information relating to the biomedical sciences, the reader is urged to carefully review and evaluate the information provided herein.

Library of Congress Cataloging-in-Publication Data

Plant biotechnology transfer to developing countries / [edited by]
David W. Altman, Kazuo N. Watanabe.
       p.cm.—(Biotechnology intelligence unit)
Includes bibliographical references and index.
ISBN 1-57059-238-1
1. Plant biotechnology—Technology transfer—Developing countries.
2. Plant biotechnology. I. Altman, David W. II. Watanabe, Kazuo N. III. Series.
SB106.B56P593 1995   95-1237
338.1'62—dc20   CIP

Publisher's Note

R.G. Landes Company publishes five book series: *Medical Intelligence Unit, Molecular Biology Intelligence Unit, Neuroscience Intelligence Unit, Tissue Engineering Intelligence Unit* and *Biotechnology Intelligence Unit.* The authors of our books are acknowledged leaders in their fields and the topics are unique. Almost without exception, no other similar books exist on these topics.

Our goal is to publish books in important and rapidly changing areas of medicine for sophisticated researchers and clinicians. To achieve this goal, we have accelerated our publishing program to conform to the fast pace in which information grows in biomedical science. Most of our books are published within 90 to 120 days of receipt of the manuscript. We would like to thank our readers for their continuing interest and welcome any comments or suggestions they may have for future books.

Deborah Muir Molsberry
Publications Director
R.G. Landes Company

A farmer in eastern Thailand plants a papaya seedling that might resist disease caused by Papaya Ringspot Virus (PRV). This project tested the use of cross-protection technology which was the basis for the development of transgenic papaya with coat protein-mediated resistance to PRV. Transgenic papaya for PRV resistance was the first genetically-engineered tropical fruit that advanced to field tests and is part of plant biotechnology transfer programs in Brazil, Thailand, and other developing countries. The program originated with Dr. Dennis Gonsalves' research group at the New York State Experiment Station, Geneva, New York, USA. Photo courtesy of D. Gonsalves.

CONTENTS

SECTION I

Section IV
REVIEWS FROM ORGANIZATIONS FACILITATING PLANT BIOTECHNOLOGY TRANSFER

IV.a   Philanthropic Foundations

EDITORS

David W. Altman
UST
Nashville, Tennessee, U.S.A.
and Department of Plant Breeding and Biometry
and Department of Agricultural
Resources and Managerial Economics
Cornell University
Ithaca, New York, U.S.A.
Chapter 2

Kazuo N. Watanabe
Department of Plant Breeding and Biometry
Cornell University
Ithaca, New York, U.S.A.
and The International Potato Center (CIP)
Lima, Peru
Chapter 13

CONTRIBUTORS

M.Y. Aziah
Forest Research Institute
  of Malaysia (FRIM)
Chapter 8

Sakarindr Bhumiratana, Director
National Center for Genetic
  Engineering and Biotechnology
Bangkok, Thailand
Chapter 9

N.C. Brady
UNDP
New York, New York, U.S.A.
Chapter 18

A.H. Bunting
University of Reading, UK
Chapter 1

S.C. Cheah
Palm Oil Research Institute
  of Malaysia (PORIM)
Chapter 8

Zhangliang Chen
College of Life Sciences
Peking University
Beijing, China
Chapter 10

K.F. Cheong
Rubber Research Institute
  of Malaysia (RRIM)
Chapter 8

Joel I. Cohen
Intermediary Biotechnology Service
International Service for National
  Agricultural Research (ISNAR)
The Hague, The Netherlands
Chapter 19

John H. Dodds
Agricultural Biotechnology
  for Sustainable Productivity
Department of Horticulture
Michigan State University
East Lansing, Michigan, U.S.A.
Chapter 17

Peter Gregory
Department of Plant Breeding
  and Biometry
Cornell University
Ithaca, New York, U.S.A.
and
The International Potato Center
Lima, Peru
Chapter 13

Hongya Gu
College of Life Sciences
Peking University
Beijing, China
Chapter 10

J. Hafsah
Rubber Research Institute
  of Malaysia (RRIM)
Chapter 8

M.D. Hassan
Malaysian Agricultural Research and
  Development Institute (MARDI)
Chapter 8

Anne Kathrine Hvoslef-Eide
Department of Horticulture
  and Crop Sciences
The Agricultural University
  of Norway
Chapter 3

Keiji Kainuma
Japan International Research Center
  for Agricultural Sciences
  (JIRCAS)
Tsukuba, Japan
Chapter 16

Muffy Koch
Innovation Biotechnology
Noordwyk, South Africa
Chapter 5

John Komen
Intermediary Biotechnology Service
International Service for National
  Agricultural Research (ISNAR)
The Hague, The Netherlands
Chapter 19

B. Krishnapillay
Forest Research Institute of Malaysia
  (FRIM)
Chapter 8

David R. Lee
Department of Agricultural,
  Resource and Managerial
  Economics
Cornell University
Ithaca, New York, U.S.A.
Chapter 4

F.C. Low
Rubber Research Institute
  of Malaysia (RRIM)
Chapter 8

Ibrahim Manwan
Central Research Institute
  for Food Crops
Bogor, Indonesia
Chapter 7

Sugiono Moeljopawiro
Central Research Institute
  for Food Crops
Bogor, Indonesia
Chapter 7

Yoshihiko Nishizawa
Japan Bioindustry Association
Tokyo, Japan
Chapter 21

K.V. Raman
International Service for the
    Acquisition of Agri–Biotech
    Applications
Cornell University
Ithaca, New York, U.S.A.
Chapter 20

Rafael Rivera-Bustamante
CINVESTAV Unidad
Irapuato, Gto., Mexico
Chapter 11

Odd Arne Rognli
Department of Biotechnological
    Sciences
The Agricultural University
    of Norway
Chapter 3

Maria José Amstalden Sampaio
CENARGEN/EMBRAPA
Brasilia, Brazil
Chapter 12

Sutat Sriwatanapongse
Deputy Director
National Center for Genetic
    Engineering and Biotechnology
Bangkok, Thailand
Chapter 9

Seizo Sumida
Japan Bioindustry Association
Tokyo, Japan
Chapter 21

M.S. Swaminathan
World Food Prize Laureate
Madras, India
Foreword

Gary H. Toenniessen
The Rockefeller Foundation
New York, New York, U.S.A.
Chapter 14

Jari P.T. Valkonen
Department of Plant Production
University of Helsinki
Viikki, Finland
Chapter 13

Bert Visser
Special Programme Biotechnology
    and Development Cooperation
Ministry of Foreign Affairs
The Netherlands
Chapter 15

John S. Wafula
Kenya Agricultural Research
    Institute
Nairobi, Kenya
Chapter 6

Jocelyn Webster
CSIR, Food Science and Technology
Pretoria, South Africa
Chapter 5

Hans Wessels
Special Programme Biotechnology
    and Development Cooperation
Ministry of Foreign Affairs
The Netherlands
Chapter 15

R. Wickneswari
Forest Research Institute of Malaysia
    (FRIM)
Chapter 8

Dr. David Altman

Dr. David Altman holds a position with UST based in Nashville, Tennessee. His educational background includes a Ph.D. (1983) in Plant Genetics from the University of Minnesota, postdoctoral training (1983) in Cytogenetics with Dr. Ron Phillips at the University of Minnesota, M.S. (1981) and B.S. (1978) degrees in Crop Science from Oregon State University, and a B.A. (1972) in French from Vanderbilt University. He also served as a Peace Corps volunteer in North Africa (1974-75). Over 90 of his scientific reports have been published. Recruited to serve as the first President and Executive Director of a non-profit international organization (ISAAA) primarily dedicated to transfer of proprietary technology to developing countries, he joined the faculty as a Professor in the Department of Plant Breeding and Biometry at Cornell University (1992-1995). As a Research Geneticist at Texas A&M University with the U.S. Department of Agriculture (1983-1992), he established an acclaimed program in cotton genetics and biotechnology focusing on insect and disease control. Among some notable achievements were preservation and use of novel chemistry from exotic germplasm, the development and characterization of tissue culture systems, and co-development of Bt transgenic plants. The annual value to U.S. agriculture of commercial products utilizing this research has been estimated at $75 million. Selected honors have been: Vice-Chair of the Faculty of Genetics of Texas A&M University, Member of the National Agricultural Biotechnology Council; Chair of the 1992 World Congress on Cell and Tissue Culture, Associate Editor of *In Vitro Cellular and Developmental Biology;* and Member of an Editorial Advisory Council of CAB International.

Dr. Kazuo N. Watanabe

Dr. Kazuo N. Watanabe is an Adjunct Assistant Professor in the Department of Plant Breeding and Biometry at Cornell University and Senior Cytogeneticist with the International Potato Center in Peru. He earned a Ph.D. in Plant Genetics at the University of Wisconsin-Madison and M.S. (Plant Breeding and Genetics) and B.S. (Horticulture and Agronomy) degrees from Kobe University, Japan. He teaches a core plant cytogenetics course and has been engaged in technology transfer of plant sciences to developing countries. Dr. Watanabe is the author of research papers and book chapters on genetics, cytogenetics, cellular and molecular genetics, breeding, and biotechnology for plants. His major emphasis is with application and transfer of plant genetic resources and biotechnology for agricultural development via crop improvement. Dr. Watanabe was honored as a Forret Forsty Hill International Agriculture Fellow at Cornell University for outstanding contribution to plant germplasm utilization and plant biotechnology transfer for international communities.

Plant biotechnology has become a priority area for technology transfer in developing countries where production of food, feed and fiber is of vital concern. Many programs now have sufficient experience to permit an in-depth examination of approaches, achievements, controversies and anticipated benefits. Developing countries are showcased for leading-edge advances, as represented by contributions from South Africa, Kenya, Indonesia, Malaysia, Thailand, China, Mexico, Brazil and Peru with a foreword from World Food Prize Laureate, M.S. Swaminathan. These presentations are augmented by reviews from organizations facilitating plant biotechnology transfer, including philanthropic foundations, bilateral and multilateral organizations and other new initiatives. Introductory chapters address the subjects of sustainable development, regulatory concerns, accessibility of resources, environmental issues and socio–economic research.

Acknowledgments

Books often are considered the bane of scientific authors because these manuscripts are time consuming and distract scientists from other publishing projects such as journal articles, the actual coin of the realm for amassing accolades. Therefore, we wish to acknowledge gratefully the effort put forth by the contributors, particularly as we were fortunate in gaining acceptance from our first choices for individuals who could best relate current state-of-the-art perspectives on plant biotechnology transfer to developing countries. Additionally, we intended to produce an international analysis, so the majority of contributors use English as a second language, and among the native speakers there were a variety of forms of the lingua franca. Our thanks to all the contributors for their patience with the process. Finally, there are several people who deserve special thanks: our spouses for assisting with international communications, Kathleen Altman and Junko Watanabe; typing services for some manuscripts, Karen Sanchez; electronic transfer assistance and correspondence, Dolores Walker; and publishing assistance from several professionals at R.G. Landes, particularly Francine Daniel in the early stages.

David N. Altman
Kazuo N. Watanabe
Ithaca, New York, USA
December 1994

FOREWORD

The human population of our planet is expected to reach about 8 billion by the year 2030. Over 7 billion of them will be in developing countries. Several eminent demographers have predicted that by the year 2030, developing countries like China and India will have to import large quantities of food to meet their food security needs. The increased demand for food will come from both a larger population and higher purchasing power. With affluence, the demand for high energy foods like meat and other animal products will also grow. Lester Brown and Hal Kane in their book "Full House" (1994) have pointed out that while demand for food will increase, the potential for meeting the demand will decrease. The adverse factors are ecological and socio-economic. The per capita availability of land and water is steadily going down, while biotic and abiotic stresses limiting crop production are increasing.

Industrialized countries which are expected to fill the gap between supply and demand in developing countries also face serious environmental problems. The threats to sustainable advances in biological productivity are growing in the industrialized world due to both high consumption of fossil fuel energy and unsustainable lifestyles. Therefore, further intensification of farming in industrialized countries will be an ecological disaster.

In contrast, a majority of the population in developing countries depend on agriculture for their livelihoods. In India, for example, over 70% of the population depend upon agriculture and allied occupations for their daily bread. Further, the problem of unemployment and under-employment is even now serious. Therefore, the import of food by predominantly agricultural countries will have the same impact as importing unemployment. Thus, while further agricultural intensification will be ecologically disastrous in industrialized countries, it will be socially disastrous in developing countries.

The green revolution of the mid 1960s and 1970s was triggered by a yield enhancing strategy. The additional production in crops like wheat and rice came from higher yield and not from greater area. The term 'green revolution' was coined in 1968 by Dr. William Gaud of the U.S. Department of Agriculture to signify this shift in the pathway of crop production. There was certainly a revolution in production and productivity in the major cereals, particularly wheat and rice. In recent years, there is evidence for the occurrence of a fatigue in the spread of green revolution. How then can we continue to improve productivity but without the adverse ecological and social consequences often associated with the older green revolution technologies?

If we are to achieve a "green green revolution" or what has been termed by some a "super green revolution," we have to seek the help of new tools of science. Among such new tools, biotechnology occupies a place of pride. The present publication provides many examples of the power of the new tools of biotechnology and particularly of recombinant DNA technology. The book also details the problems and potential in the transfer of these new technologies to the developing countries. On the one hand, developing countries require the latest tools which science can offer to improve the quality of life of the rural and urban poor and on the other hand, these technologies require for their adoption several substrate requirements including the availability of highly skilled technical personnel. Before I go into the methods by which we can achieve the sharing of the benefits of new scientific findings, I would like to refer to some of the major challenges.

First, the population of the Third World will increase by 100 million more people every year, the largest annual population increase in history. This growth, combined with increased demand caused by rising income, will lead to large increases in food demand.

Second, growth in incomes and urbanization is expected to increase consumer demand for livestock products, which in turn will increase the demand for feed grains.

Third, the increasing demand for food and feed must be met primarily from land already under cultivation; in fact it will need political will and action to preserve prime farm land from being diverted to non-farm uses.

Fourth, employment opportunities for the rapidly expanding labor force will have to be found primarily in the food and agriculture sector and in those non-farm sectoral activities that accompany agricultural growth. The livelihoods of the rural poor are particularly dependent on this growth.

Compounding all these problems is the growing complacency about the ability of the world to meet future demand and increase the consumption capacity of the poor. Current world surpluses and consequent low prices have tended to depress returns and discourage investments in agriculture. The growing obsession only with policies for industrial growth overlooks the fact that the prospects of non-agricultural or overall development are linked with agricultural progress. How then do we promote a job-led economic growth strategy based on the principles of ecology, equity and employment?

Sir Francis Bacon once said:

"It would be an unsound fancy to expect that things which have never yet been done can be done except by methods which have never been tried."

It is in this context that we have to review the role of biotechnology in fighting the fatigue of the green revolution. Green revolution, which is another name for enhancing crop production through productivity improvement, or what I have been calling "land and forest saving agriculture" is an ecological and economic necessity in the Asia-Pacific region. Sustaining and expanding the green revolution is an ecological necessity, as otherwise the remaining forest land will be invaded for crop cultivation. It is an economic necessity because only a vertical growth in productivity can help to cut costs and increase income from small holdings. Most of Asia's farmers are small cultivators. The next stage in the green revolution should however take note of the ecological problems associated with genetic homogeneity, soil degradation, excessive use of mineral fertilizers and chemical pesticides and the unsustainable exploitation of ground water. This is where the suggestions contained in chapter 16 of Agenda 21 of the U.N. Conference on Environment and Development relating to environmentally sound management of biotechnology assume importance.

The feedstock for the biotechnology industry is biological diversity. The Global Convention on Biological Diversity which became operational on 29 December 1993, recognizes in Article 16 the importance of both access to and transfer of biotechnology in achieving the objectives of the convention in the areas of genetic conservation, evaluation and utilization.

The advent of biotechnology has opened up new opportunities for crop, tree, animal and fish improvement, by rendering novel genetic combinations feasible. It has generated new opportunities for skilled employment in villages through biological software industries, such as the manufacture of biofertilizers and biopesticides and establishment of biomass refineries. Hence, the Biodiversity Convention calls for both sustainable and equitable use of biological diversity.

At the same time, the advent of biotechnology has created a potentially dangerous situation in the pattern of international cooperation in science and technology. The World Trade Agreement signed on 15 December 1993 by the nations adhering to the GATT agreement, makes patenting or some appropriate form of varietal protection mandatory. The systems of plant variety protection so far in operation in industrialized countries under the UPOV Convention recognize and reward only the breeders. In the forum of FAO, the concept of Farmers' Rights was developed during the last decade to stress the need for also recognizing and rewarding the contributions of rural and tribal women and men who have over the millennia not only conserved plant and animal genetic resources but have also practiced selection for resistance to biotic and abiotic stresses and for medicinal and other economic properties. At a recent inter-disciplinary dialogue held at Madras, a Plant Variety Recognition and Protection Act has been developed, which provides a mechanism for integrating equity and ethical considerations in the sharing of profits from plant breeding research involving both conventional and molecular approaches.

Nearly 70 years ago, JBS Haldane in his book 'Daedalus' or 'Science and Future' (1923) urged that unless our ethical outlook evolves to keep pace with development of science and technology, social chaos and disaster would result. In this context some recent developments are disturbing. Examples include patenting by commercial companies in industrialized countries of Margosan-O developed from the Neem tree, an extract from the African soapberry plant Endod used for killing zebra mussels, an anti-cancer extract from the Indian medicinal plant *Camptothescin* and to cap it all, a broad patent which could cover all genetically engineered cotton. These events are leading some countries to take pre-emptive measures which will prevent others from filing patent claims. For example, the state of Queensland in Australia, has passed legislation giving it intellectual property rights over genetic information embodied in all the plants and animals found in the state. Similar initiatives are yet to be taken by developing countries, although such pre-emptive strikes are likely to be taken soon by other countries to prevent a few cashing in on the silent work over centuries of nature and human families.

Action is urgent particularly in the field of medicinal and aromatic plants in which the Asia-Pacific region is rich, since the additional trade in medicinal plant products is likely to grow by billions of dollars before the end of the century.

Let me cite two examples from recent research in India to give a glimpse of the untapped potential. First, the compound piperine extracted from the fruits of *Piper longum* helps in the absorption of antibiotics such as rifampicine. Tuberculosis patients can get cured with a lower dose of rifampicine when it is taken in combination with piperine.

Second, baccosides extracted from the brahml plant, *Baccopa miniori* help to improve memory among the aged. Fortunately, this compound has been patented by the Central Drug Research Institute at Lucknow, India.

There are numerous opportunities of this kind which the biotechnologists of our region should seize, as otherwise the present situation where genetic richness and economic impoverishment co-exist will continue.

We should take note in our research work of three other major concerns.

First, the need to preserve genetic heterogeneity in cultivated plants, trees and animals, in order to avoid genetic vulnerability to pests and diseases.

Second, the importance of a proactive analysis of the potential impact of transgenic plants and micro-organisms on the environment.

Third, the need to enhance public understanding of the real implications of recombinant DNA-technology. We cannot deny that there is a growing public distrust of anything that smacks of genetic engineering. For example, the U.S. Food and Drug Administration has recently taken the extraordinary step of warning dairy farmers that they should not label milk as "hormone free" unless the milk is from cows not fed with bovine somatotropin (bST).

Such apprehensions should not be dismissed as arising from ignorance, since they are rooted in experience where scientists have often failed to take a holistic view of all the likely effects of new technologies. We need a well organized movement for promoting public understanding of biotechnology. Social scientists and biotechnologists should work together from the time a project idea is conceived, so that technologies emerging from expensive research become socially acceptable. The Rockefeller Foundation sponsored Rice Biotechnology Network shows how this can be accomplished.

U.N. organizations like FAO, UNEP, UNESCO, UNIDO, WHO and IAEA have taken several steps to assist in spreading the benefits of new biotechnologies. The Asian biotechnology and biodiversity sub-programme of the Farmer-Centred Resource Management (FARM) programme of FAO is an example. UNEP is actively following up the implementation aspects of the Biodiversity Convention including the establishment of a Biodiversity Trust Fund. The International Centre for Genetic Engineering and Biotechnology, with its twin Centres in Trieste and New Delhi, so far nurtured by UNIDO has now become an autonomous international research institution with its own Board of Trustees. Non-governmental initiatives like ISAAA (International Service for the Acquisition of Agro-biotech Applications) are also timely.

While all these international initiatives are welcome, the primary responsibility for strengthening national capability in genetic enhancement through molecular techniques rests with the government, universities and research institutions of each country.

It is my hope that ISAAA will stimulate and assist the growth of purposeful partnerships among universities and research institutions in the industrialized countries and the research institutions in the developing countries. Private sector industry also can play a critical role in poverty alleviation and in promoting sustainable livelihoods among the rural and urban poor. We need new forms of social contracts between private sector industry and resource poor families. There is evidence that development which is not equitable will not be sustainable in the long term. This is why I salute the editors and authors of this book for their commitment to the cause of harnessing the power of modern biotechnology for strengthening national and global food and nutrition security systems. I hope this book will be read widely.

Prof. M.S. Swaminathan

INTRODUCTION AND OVERVIEW PERSPECTIVES

CHAPTER 1

PLANT BIOTECHNOLOGY AND DEVELOPMENT: A BEGINNER'S GUIDE

A.H. Bunting

The title of this book suggests that scientists and technologists working in developing countries need more plant biotechnology to help biological producers to manage, conserve and use natural resources more productively in support of development. Long experience of research intended to lead to increases in biological output tells us that science and technology alone are not enough to promote development. This chapter is about the enabling conditions, outside the fields of science and technology, which have to be sufficiently met if new knowledge is to help to advance development in this way.

A paragraph so replete with abstract nouns invites definitions, with which, therefore, we begin. In what follows, *biological products* include food, fiber, fuel and timber, amenity and medicinal plants; livestock including birds, and the plant products on which they feed; fish and other aquatic products; and microorganisms. All these products depend ultimately on energy fixed and materials elaborated in green plants; and they use or compete for the same or similar sets of natural resources.

DEVELOPMENT

DEVELOPED AND DEVELOPING NATIONS

We were all undeveloped once. Today, the nations of the earth form a continuous spectrum on the road of development (Food and Agriculture Organization of the United Nations;[1] World Bank,[2] which tabulates many quantitative indicators of development). We can define development by its works, from the differences along this spectrum.

The more *developed nations* include about 50 of the richer countries of the world[4]—the United States and Canada, and the nations of Western and Eastern Europe and of part of the former USSR, together with Australia, New Zealand, Japan, Israel and South Africa. In 1995 they contained about 1.3 billion people out of a total world population of about 5.7 billion. Most of the inhabitants of these very diverse nations are better fed, housed, clothed and educated than those of other nations; they are more healthy, they live longer and their average incomes are larger.

The 170 or so *developing nations* are the less prosperous countries, including the poorest on earth. They comprise all the remaining nations of Asia, Africa, the Americas and the Pacific. In 1995 they contained 4.4 billion people, and they are poorer than the developed countries in respect of all World Bank indicators.[2] They too are very diverse, in size, environments, resource endowments, density and rate of growth of population, diversification of economies, degree of development, and international historic and political associations.

They are especially diverse in their command of science and technology. Thus, for example, molecular biology is comparatively strong in China and in Argentina and some other nations of Latin America, but appears to be weak, at best, in several nations of Africa.

Though it is consequently unsafe to generalize about "the developing countries" or their needs, they do share two characteristics which are particularly important for our present purpose. The first is geographical and environmental. Most of them lie between the northern and southern desert belts of the earth. As a result their seasonal weather regimes are not simple extensions of those of the temperate or winter-rainfall regimes on the poleward sides of the desert belts: they are mirror images of them.[3] The Western Sudan is not a hotter version of Nebraska. The seasonal courses of water balance are totally different. The consequences for agronomic potential and production methods are profound, even if not all visiting experts perceive them.

Second, most of the developing nations are eligible for development assistance. This often includes support for scientific and technical cooperation with institutions in developed nations, and so provides a conduit along which new knowledge can reach and be implanted in the institutions of the less-developed countries.

DEVELOPMENT AND ECONOMIC DIVERSITY

The historic process which we call development began about ten thousand years ago with changes in biological production. At that period, the hunting and gathering of wild species, to meet immediate needs, began to give way to organized management, production and harvesting, centered on settlements. Seasonal surpluses required the post-harvest treatment and storage or preservation of the products.

But biological production did not stand on its own. The skills of the self-sufficient household were soon supplemented by division of labor and the emergence of more specialized sectors, including the spinning and weaving of fibers, and the mining and working of wood and metal, and so of manufacturing and other industries, as well as of an ever-increasing volume of services of many sorts. As populations increased and markets developed into towns and cities, the economics of societies not only increased in size: they became vastly more diverse.

In the modern world, the nonbiological sectors provide the main market for biological output, surplus to the subsistence requirements of the producers, and are deliberately produced for sale. The biological producers and their endeavors do not form a separate self-contained sector of national and international economies, developing in isolation: they are intimately linked with all other sectors, at home and abroad, on which they depend not only for their markets but also for domestic and consumer goods and production inputs.

Except perhaps in some of the few remaining subsistence societies, biological production is a business, in which the producers seek to maximize the return to themselves on their investment of resources and the risks they accept. Modern agriculture (along with other forms of biological production) has evolved as the agriculture of cities, which purchase the surpluses and manufacture many of the inputs necessary to sustain the output and maintain the profits of biological production. In the more-developed countries, few farmers produce their own staple foods. Most rural people buy their clothes in the stores of the city.

A DEFINITION OF DEVELOPMENT

From the observed contrasts and differences in this changing and evolving scene we may now attempt to define development. Historically, development has been an uneven process of linked technical, economic, social and political change, in which the resources commanded by social groups (from family and household to nation, confederation and union) are made to produce larger quantities of an increasingly diverse range of goods and services which the groups, or individuals and subgroups within them, desire. The creation of wealth in ever larger total quantity, and in ever more forms, has been a central feature.

Along the road the structures of economies change irreversibly. At one time or another, all of us were rural subsistence producers. In 1992, only 8% of the economically active population in the developed nations were engaged in biological production (compared with 15% in 1975). In the developing countries the corresponding numbers were 58% in 1992 against 68% in 1975, in spite of substantial increases in the absolute sizes of their populations. Development does not preserve rural economies in their historic molds: it transforms them.

DEVELOPMENT AND EQUITY

Development has always been driven by self-interest, greed, and the desire for what people regard as progressive change for themselves—more wealth, more land, more food, more power, a less-laborious life. It has often been marked by invasion, conquest, subjugation, slavery, revolution and other forms of violent and oppressive change.

Even if it has seemed peaceful, development has not been inherently pleasant, benevolent or beneficial to all who have been affected by it. The distribution of the proceeds of development has depended on the pattern of control, within the social group, over resources and products. Where this was or is communal, and at least partly democratically determined (as in the former "three fields" system in England and in the customary land-holding systems in many African societies), the outcome may be at least partly equitable. But otherwise, it is all too evident that development has usually increased inequity. Few of the most highly-developed nations in the world today are equitable in their social structures. It is not inherent in the development process that it lessens the poverty and misfortune of the poorest. Its outcome has been equitable only to the extent that political and economic pressures have compelled it to be so.

DEVELOPMENT AND THE ENVIRONMENT

Development has produced substantial change in the natural environment. It began alongside the retreat of the ice, in the continuing global warming which has followed the last glacial period. It has produced or accelerated significant changes in the terrain and in virtually all of the wild vegetation of the earth. Along the way, marginal areas have been abandoned to dereliction, as alternative ways of making better livings have emerged. The secondary forest of much of the New England states is a well-known example. Indeed, the longer-term future of biological production and of rural producers in the developing countries does not lie in the marginal areas: it lies increasingly, as it has done in all developed nations in the world in the past, in the more favorable areas, where output and profit can be increased more readily and more securely, and costs per unit of product are smaller.

THE DEVELOPMENT DEBATE

These perceptions of inequity and environmental change associated with development have led to vigorous debate among development practitioners and observers, particularly those who live in developed countries. This post- or neo-colonial debate is not so much descriptive, about what development has been in practice, as normative, about what different people think development ought to be. Is it enough that development should simply increase the total output of goods and services, with little reference to the distribution of the proceeds or change in the environment, or can it, and if so should it, be designed so as to ensure equitable distribution, the relief of poverty, and the social changes desired by different sorts of observers, together with sustainability (whatever that may be), conservation and a range of other environmental desiderata?

Regardless of the shrill background noise of this largely ideological debate, development lurches onward. Intellectual and political forces, even when they are continued into war, have surprisingly little effect on its uneven, but seemingly inevitable course.

DEVELOPMENT AND BIOLOGICAL PRODUCTION

All human societies require biological products. I have suggested above that development started with the agricultural revolutions which began, apparently separately in several different regions of the earth, about ten millenia ago. Since then, along with the increase in human numbers, the economic biological output has been enormously increased. In most environments, biological productivity has not merely been sustained—a modest ambition—it has been greatly increased. The increases rest on larger outputs of plant biomass coupled with increases in the proportions of the biomass that can be utilized for economic, aesthetic or medicinal purposes.

Throughout its long history, one of the main tasks of the evolving science of plants has been to strengthen the knowledge base for these advances in biological production. Whatever else it has done, botany has always contributed to economic progress.

PLANT BIOTECHNOLOGY

At this point we require our last definition—that of plant biotechnology itself. The words are defined more rigorously elsewhere in this book. But as I hear them in everyday chatter, they constitute a promotional, Humpty Dumpty term, generally used to mean exactly what each speaker intends it to mean, neither more nor less. Nonprofessionals of all sorts flaunt it to display their scientific erudition.

Plant biotechnology appears to include a variety of topics old and new, including fermentations; stem-tip, tissue and cell culture and related methods of regeneration of entire plants; analytical and diagnostic methods based on biological or physico-chemical separations or biological recognition phenomena at the molecular level; and (most important of all) new and rapidly evolving ways to modify the heritable attributes of plants which are based on the manipulation, management, modification and fabrication of sequences of nucleotides. The term often refers to applications rather than to new science.

PLANT BIOTECHNOLOGY, BOTANY AND DEVELOPMENT

The scientific endeavors which sustain plant biotechnology are seen by many as the sharpest and most exciting leading edges of botany at the present time. But from the point of view of development, plant biotechnology is no different in principle from the wide range of older uses of botany which mankind has developed during the past ten thousand years to describe, classify, improve, propagate, raise, protect and use plants. This is the original historic applied stock of botany, from which sprang the more modern formal shoots of "pure" botany, including morphology and anatomy, plant physiology and biochemistry, plant pathology and genetics, all of which in their turn have helped to strengthen the ancestral applied stock.

All the uses, and much of the science, of plant biotechnology rest on these older fields of knowledge. Plant biotechnology can advance and improve classic areas of botany, but it does not displace them. So one must regret that some competent practitioners at the molecular level appear to have little interest in more old-fashioned topics such as the morphology and ontogeny of plant structure and the eco-physiology of adaptation, or in reproductive biology, which are so important for both agronomy and crop improvement in the real world.

PAST ACHIEVEMENTS AND FUTURE TASKS IN BIOLOGICAL PRODUCTION

I now turn to the practical achievements and future tasks of biological production. In these paragraphs, the output of cereals, the most important foodstuffs of mankind, is used as a proxy to indicate the important trends of the past and to outline the tasks ahead. The output of other commodities has changed similarly, and may be expected to change even more rapidly in the future.

THE ACHIEVEMENTS OF THE PAST

During the past sixty years, as the human population of the earth has very nearly trebled, from about 2.0 billion to 5.7 billion, the output of cereals per year has increased from about 600 million tons to around 1,950 million. This represents an average increase in output per head from about 300 to around 345 kg per year. To be sure, the increases have been proportionately larger in the richer than in the poorer nations, but nevertheless the average for the developing nations has increased from about 220 to 250 kg per head per year even though population has increased more than threefold.

In both developed and developing countries, the increased output has come mainly from substantial increases in yield (+171% and +119% respectively). In developed countries, harvested area has actually decreased by about 10%, while in developing countries it has increased by 60% only.

The increased output has been produced in response to increases in effective market demand, generated by the growth of human numbers and by development in other sectors and in other places. These increases have been accompanied by improvements in average nutritional intake[1] and in average life expectancy at birth (among many other indicators).[2,4]

THE TASKS OF THE FUTURE

This section sets out a crude assessment of the "engineering tasks" for applied biology during the century ahead.

Population

The average growth rate of world population reached a peak at 2.3% per year around 1960. Since then it has steadily decreased, to about 1.5% per year in 1995. In the half-decade of 1990-95, we reached the historic peak of absolute growth rate of the human population, at about 85 million per year. Henceforth the annual increments are expected to be progressively smaller. Nevertheless the total number of humans is expected to double through the coming century, as both birth and death rates continue to fall and average life expectancy increases, to reach a more or less "plateau" level at about 11 billion.[4] The largest proportionate increases are expected in west Asia and in Africa both north and south of the Sahara. AIDS notwithstanding, populations in these regions seem likely to quadruple during the coming century.

Effective demand

Through the century ahead, as both human numbers and average purchasing power continue to grow, we may expect that the effective demand for biological products will continue to increase. So, for example, the total requirement for cereals in the year 2095 may be between 5 and 6 billion tons instead of nearly 2 billion now. Most of this extra demand will arise in the presently developing nations, as a continuing part of their normal development.

It has taken our species 10,000 years to raise biological output to its present level. In the period of 100 years only which lies ahead, the world's biological producers appear likely to double or treble the present output. They will probably achieve, in one century, an increase in output our ancestors produced in 100 centuries. This is the measure of the future task for biological producers and for the botanists and other scientists of all kinds who support them.

Yield

Forward projections from the trends of the past suggest that the harvested area of cereals in the world may increase from its present level of about 0.75 billion hectares to about 1.0 billion. To produce 5 or 6 billion tons from this area will require average yields of 5 or 6 tons per hectare, against 2.8 tons per hectare in 1992 (3.2 and 2.5 tons per hectare in developed and developing countries respectively). The present levels of yield are limited, not by environmental considerations so much as by profitability and avoidance of risk. Many observers, and most agronomists, seem to forget that farmers are interested in profit and security rather than in yield. No producer, even if he is illiterate, will pursue a diminishing returns curve to the bitter end.

The FAO study of agro-ecological zones and population-supporting capacity, cited below, the existing range of recorded yields of commodities in different nations and environments, of the experience of experimenters, and the achievements of the most skilled practical farmers, suggest that on average, the necessary average levels of yield can be realized as it becomes profitable to do so.

Environmental resources

Most of the extra output will be produced in the developing nations. The environmental (agro-ecological) resources of these nations for biological production are sufficient, in total, to sustain the expected populations, given some technically feasible improvements in production methods, but they are unevenly distributed.[5]

Some nations, for example those of North Africa and West Asia, already contain, or will soon contain, more people than their environments can sustain. But others, in all continents, are potential producers of surpluses. The environmental resources of the 24 potential surplus producing nations of Africa south of the Sahara are amply sufficient to support the expected additional populations of the 18 potential deficit nations, and to produce a large surplus for export to other regions into the bargain. As output increases, both vegetation and terrain will be changed, as in the past.

Tasks for biological producers and industries

The future task for the biological producers in the developing nations, and their immediate customers, the merchants and post-harvest industrialists, will therefore be to produce, market, store, process and transport this rapidly and substantially-increasing tonnage of output, and to deliver it to the final users and consumers. The global trade in cereals, at present reported at 190 million tons per year (about 10% of total output), may increase to 500 million tons or more. More and more of it will be between countries which are at present regarded as less-developed. Against this prospect, I cannot but regret that so many devoted development practitioners, preoccupied close to the grass roots with the urgent human needs of the present in poor rural communities, seem to be unprepared for the vast increases in output which will be required in the future—which, in the present half-decade of peak absolute increments of human numbers, is already upon us.

SEVEN GROUPS OF RESOURCES AND CONSTRAINTS FOR BIOLOGICAL PRODUCTION

HANDLER'S LAWS

My starting point here is what their inventor, the late Professor Philip Handler, called Handler's laws about food.[6] After pointing out that ample food was already produced in the world, and that it is the poor who go hungry, he reminded us that "nowhere in the world are farmers happy about growing more food than they and their families can eat, unless someone gives them something they want for the extra food they produce." Put another way, it is the prospect of profitable sales off the farm that evokes extra output, not only of food but of all biological products, and may encourage producers to use new techniques to get it.

No offering of new technology is likely to be adopted, even if it does what it is cracked up to do, unless it helps producers to get a larger return on the resources they commit than they could achieve, without undue risk, by using those resources in other ways. This is one of the central reasons why innovations which are technically satisfactory on the research station are so often of little interest to practical producers. There is not enough profit in them.

Moreover, knowledge, whether of plant biotechnology or of any other sector of research and technology, is far from the only factor which affects the development process. Knowledge, and especially new knowledge produced by research, may be (though it is not always) necessary for development, but it is seldom or never sufficient. Knowledge does not of itself lead producers to deliver increased output. Research is not the engine of development.

Studies and experiences of change in biological production (for earlier examples see Bunting,[7] Hunter et al[8] and Bunting[9,10]) have suggested a checklist of the main groups of factors which may support or constrain increases in output. There are seven such groups. Output increases only if the circumstances are sufficiently favorable in all of them. Weakness in any of them tends to restrict output.

INTERNAL POLICIES AND PRACTICES OF GOVERNMENT

Governments seldom make good farmers. Though there have been exceptions, direct economic actions by governments in production, post-harvest treatment and processing, storage and distribution have generally failed signally, and often at vast expense, in both developed and developing countries.

The main task of government in relation to biological production is to establish and maintain an *enabling environment* in which it can develop. Among the components of this environment are the maintenance of peace, law and order; competent management of the economy and its foreign relations including the state budget and the real exchange value of the currency; elimination and prevention of corruption, particularly in the public service; monitoring and projecting the growth and distribution of population; forming an inventory of natural resources for development (including water and minerals as well as terrain, soils, weather and climate); forming from it an evolving policy and plan (including necessary physical infrastructure and power resources) for using both natural and human resources to advance the economy as a whole; establishing effective means of executing and monitoring the plan; promoting economic growth and diversification, both rural and nonrural, including mining and industrial development, in sensible balance with development in biological production; for this purpose, guiding the movement of population from rural areas, particularly those with more marginal environments for biological production, to new employment in villages, towns and cities, for which both private-sector investment and training will be needed; and equipping the countryside physically to support the marketing of rural products and the development of industries, including those based on biological primary products.

It is also a duty of government to teach the people, by example as well as precept, that money is not a wild biological product which grows on trees.

We expect the state to provide services of these sorts in developed countries: they are even more necessary in the poorer and less-developed countries. This is work for professionals, in both the public service and the representative institutions; and appropriate professional training is required. A field-marshal, promoted but yesterday from the rank of lance-corporal or even sergeant, cannot be expected to command the necessary competence by the unguided light of nature.

FOREIGN RELATIONS AND OTHER EXTERNAL POLICIES

The foreign relations group includes the management of relations with cooperating nations and international public institutions, foreign exchange, development aid and indebtedness, which are tasks of government; but it also includes relations in the private sector between both state and private-sector agencies and the private-sector transnationals—examples are finance and investment groups like international banks, the international companies in the petroleum and chemical industries, civil and mechanical engineering consultants and contractors.

EFFECTIVE DEMAND

As we have noted above, without effective demand at an acceptable price in an accessible market, no surplus output will be produced. Those redoubtable private entrepreneurs, the legendary small-scale, conservative, resource-poor, risk-averse "farmers" in developing countries, allegedly embracing backward-sloping response curves, are not in the business of producing free lunches for city folks, even though governments may try to compel them to do so in order to placate the voters and potential assassins in the capital.

The bulk of the effective demand can only come, in most developing nations today, from the increasing number of customers who depend on the nonbiological sectors of a growing and diverse economy. Increases in the output of biological products respond to increases in the output and returns of other sectors, and so of the demand from those who work in them. It follows that to advance biological production, it is necessary to develop profitable nonbiological industries and services in a diverse economy. Governments have in the past been berated by the agencies for not spending enough on agriculture: seen from the mid-1990s, their fault was rather that they did not spend enough money to promote nonagriculture.

In the past, significant effective demand often arose in export markets in neighboring or more distant countries. Canada and Denmark developed advanced rural structures in order to export to Western European markets. Uganda developed in earlier times by exporting cotton and later coffee, and Zimbabwe's more recent development has depended largely on exports of tobacco. In these cases, it might have been claimed that development was led by agriculture, but in fact it was led by demand arising from nonagriculture in the importing countries.

In 1995 competition between developing nations has made the traditional (but now usually inelastic) export markets both more limited and less certain. To be sure, new export markets for biological products are already growing in potential deficit nations in the developing world itself, as population presses increasingly on environmental resources. It does not, however, seem that the new political, physical and management infrastuctures necessary for such exchanges are in prospect at the present time.

OUTPUT DELIVERY SYSTEMS

The output delivery system includes all the physical, financial and managerial facilities that are needed to move biological products from the metaphorical farm-gate to the end-user or consumer. The countryside has to be kitted-up for development.

The most important element is the physical logistic infrastructure on which biological products can be moved—bridges, roads, railways, navigable rivers, canals, airfields; the tractors, locomotives, vehicles, boats and aircraft, and their maintenance and fuel, which take the place in more-sophisticated economies of the innumerable bullock carts on which rural India goes to market. Given these physical means, development has usually proceeded by private initiatives.

Output systems also include the markets themselves, where the primary products first change hands and money moves into the rural economy; the market masters who keep track of quantities, qualities, transactions and prices; the merchants who purchase and transport the products; the domestic and commercial or official storage installations where the products are held until they are required, and the processing installations in which the primary products are prepared for final use or consumption.

Storage is necessary to manage the effects of seasonality within each year, and of fluctuations from year to year in the weather and in market demand. In customary populations of many crop species, special physical and chemical adaptations to minimize deterioration in customary storage have been found. Not all plant breeders seem to know about them, and have consequently released allegedly improved crop populations which cannot safely be stored in customary stores.

Producers may take advantage of the large potential yields and official procurement prices of such populations to raise as large an area as they can of them, so as to make a useful profit, leaving the consequent difficulties of storage for the account of the state procurement agency. Meantime they grow and store their customary varieties as usual.

Storage is intimately linked with processing—cleaning, grading, parboiling, grinding, fermenting, distilling, ginning (followed by spinning and weaving), tanning, cooking, canning—the list of prehistoric, historic and modern post-harvest processes is endless. The physical and chemical attributes required for domestic processing may not be the same as those needed by industrial-scale processors—another potential snare for the breeder and the plant biotechnologist.

Long-range transport may call for increased mechanical strength in the produce, to minimize crushing in trucks and railway wagons.

In respect of plant breeding, this section may serve to remind us that however important a single attribute may be, it is not likely to succeed (as every cotton breeder knows well) unless the population which contains it is also satisfactory in respect of the whole range of attributes valued by producers and their customers.

The broader purpose of this section is to remind us that producers can only respond to effective demand if the output delivery system is in sufficiently good shape. Surplus produce rots by the roadside if it cannot be got to market in time.

For the plant biotechnologist, this section offers opportunities in storage, protection from losses in storage, in processing, and in the physical, nutritional and aesthetic qualities of their products.

RESOURCES AVAILABLE FOR EXTRA OUTPUT

To produce extra output to meet effective demand, producers may need more land, labor and capital; more water and more farm chemicals; and other production inputs. These may include planting material adapted to their environments and production systems, as well as to the needs of their households and customers. They may also need extra power, from animals or engines, and improved equipment.

Additional resources may be used in unexpected but more profitable ways. So for example fertilizers and protection chemicals provided (along with seed, equipment and extension services) were not used in some parts of northern Nigeria to increase the output of sorghum and millet, as had been intended. They were used instead to produce more profitable products including maize for human food, and for feed for intensively raised poultry, in and around the cities of the south.

Technical support is best guided, not solely by what the local administrator or the visiting aid agent thinks producers ought to do, but at least partly by what they themselves are already trying to do.

TECHNICAL METHODS AND PRODUCTION SYSTEMS TO ACHIEVE EXTRA OUTPUT

If government policy and external relations are satisfactory, effective demand and price are satisfactory, the output delivery system is in sufficiently good shape, and resources are available to support extra output, producers may be attracted by the prospect of producing and marketing extra output. At this point many observers expect them to need, and readily to adopt, a prefabricated set of new technologies, chosen by other observers, few or none of whom are producers, and gift-wrapped as a package of practices to facilitate transfer, and which the extension service stands eagerly ready to deliver.

The observers have all too often been disappointed, for three main reasons. First, the producers, who have long experience of environments and economic organisms, may already know perfectly well how to obtain more output from the resources they command. This happened in northern Nigeria 30 years ago, when the first all-weather road was completed from Kano to Zaria. Tens of thousands of profitable hectares of chewing cane sprang up in the bottom lands along the road, and a steadily increasing number of villages became market places for fruit and vegetables. Indeed the road as a whole, all 120 miles of it, became a market place, towards which some villages moved their locations and along which truckloads of eager customers traveled day and night. In formal terminology, the limiting element had been, not the effective demand, the resources or the technology, but the output delivery system. When the new road brought the market to the doorstep, output exploded, using methods and resources already well known.

Second, the packages of practices offered have not always been acceptable to producers. Often, they have been designed to obtain maximum *yield* from a single crop, whereas the producers are more interested in maximizing the total *return* on all resources committed in all the enterprises that form the business part of their life systems.

Third, the technology offered has all too often been inappropriate for its purpose in the real-life situation of the producers. So for example, in Northern Nigeria, where the wet season varies substantially in length and total precipitation from year to year, attempts to substitute uniform, dwarf, nonphotoperiodic, short-season and allegedly drought-resistant and high-yielding "modern" sorghum for the diverse, often tall, photoperiodic customary forms failed.

Not only did the allegedly improved varieties yield less stover (useful for feed and structures), but they lacked the inbuilt flexible adaptations which enabled the customary populations not only to be drought-resistant (and so produce a moderate yield) in dry years, but also to be able to take advantage of longer and wetter seasons to produce bumper crops of insect-resistant grain to fill the domestic over-year stores, on which the people depended for survival.

The goals of research which seeks more profitable methods for producers have to be determined, not by trying to adapt systems and their components which succeed in quite different environments back home, but through analytical understanding of the purposes and methods of the existing production systems, as J.A. Voelcker[11] suggested in India more than a century ago. We may hope that as research on systems of farming (the so-called farming systems research) escapes from its present micro-economic and sociological captivity, and returns to its older roots in environmental science, agronomy and anthropology (see, for example, Allan[12]), technically-competent agricultural scientists will be able to use it to specify objectives for both development and biological research.

Where technical understanding has been adequate, as it seems to have been for the production of more productive kinds of cereals in Mexico, India, China and other Asian nations (mostly with irrigation), and in Zimbabwe and a few other nations of Africa, the results have been impressive. The job can evidently be done provided the technical tasks, as well as the social and economic elements, are correctly identified and analyzed.

This section offers opportunities for applied botany in plant improvement, protection from field pests and diseases, and propagation by seed and vegetative means. The technical interpretation of existing systems, and the selection of components for improved production systems, are customary areas for applied botanists, though they may have less to offer to the biotechnologist.

But we do not have much time. As we have noted, the world's biological producers will be called on, during the coming century, to double or treble the output of their products. The tasks ahead in using natural resources correctly and sustainably will place vast demands on the seventh group of factors, in the knowledge systems for biological production.

KNOWLEDGE SYSTEMS

Knowledge consists of factual information, concepts, techniques and skills. It is stored, increased, developed for use, disseminated and applied in knowledge systems. In our time, in spite of difficulties of language, a global, interactive and instantaneous knowledge system (largely based on the kinds of English that are understood by computers) is emerging, which links all participants who have access to appropriate facilities for information management.

In most developing nations, the main repositories of technical knowledge of environments and economic organisms are in the minds and memories of rural people. Little of this knowledge has been assembled, interpreted, codified and collated with more formal knowledge into scholarly monographs. Maybe it is thought to lack the prestige of information derived from research of western type.

The national knowledge systems which support biological production are generally both incomplete and divided between several Ministries of government. Moreover, few of their components, even within the state service, have direct responsibilities or links to the national processes for forming and implementing development plans, or to the producers, processors or ultimate users and consumers of biological products.

The Universities are usually formally separated, under a different Ministry, not only from the Ministries responsible for biological production, but also from the central, regional and peripheral agencies which plan and implement development actions. In many countries they are also separated from the public-sector research and extension systems, which may in addition be separate from each other. The Universities are often unaccustomed to cooperation with the knowledge institutions in the private sector, for example in the large commercial companies.

The remarkable national cooperative knowledge system of the United States for biological production and the management of natural resources is often cited as the shining example which it is. But it has no parallel anywhere in the world. One of the reasons why the system works in the United States, but not in all states of India (for example), is that the individual States of the Union do not have executive State Departments of Agriculture, so that the agricultural universities have sole responsibility for all publicly-funded knowledge services for biological producers and their customers.

Moreover, few nations have suitable text and training materials prepared for their environmental, economic and social conditions, even though, in many, both research and education related to biological production have been carried on for a century and often longer. Much teaching, even of environmental sciences, uses books, concepts and methods derived more or less directly from the conditions of nations in the temperate zone.

In these circumstances, it is often very difficult to determine what sorts of support plant biotechnology, or indeed any other field of applied biology, can best offer to support development in biological production, or even how to deliver what is offered.

At an even more fundamental level, though it has certainly been widely feasible to introduce into the technical services of developing nations many techniques and skills based on the newer biology (for example stem-tip culture or ELISA), in many instances weaknesses in chemistry and biological science hold back the rate at which local workers come to participate in the advancement of the basic sciences, on which plant biotechnology rests.

SOME CONCLUSIONS

Clearly those developing nations which wish it should be encouraged, and if necessary assisted, to use, and to advance for their own purposes, all branches of pure and applied science and technology which they believe will assist them in development. But the foreign cooperator, and his colleagues in the developing nation, must understand the nature and purposes of the development process in each nation, and accept that science and technology form but one of many elements that influence progress.

Moreover, the goals and objectives of science and technology have to be selected wisely. To do this for biological production the researcher and technologist must understand technically how the existing production systems have worked in the past, how they work now, and what will be required of them in the future. To achieve this, those who study the systems of production and the knowledge and experience of the producers must be scientifically and technically literate and perceptive in describing and interpreting the processes producers use.

It then becomes possible to establish quantitative objectives, in terms of the required volumes of output of products, for applied biological science and technology. These are derived from the opportunities and purposes of producers, from national, regional and local plans, from the natural resources available for biological production and from the relevant customary as well as modern scientific and technical knowledge.

Science and technology, including plant biotechnology, are certainly essential for the future progress of the developing nations. But they offer few quick fixes. Above all, they cannot offer a cheap substitute for a technically and economically sound, and a socially and administratively practicable, development policy.

REFERENCES

1. Food and Agriculture Organization of the United Nations. FAO Production Yearbook 46. Rome: FAO, 1993.

2. World Bank. World Development Report 1994. New York: Oxford University Press, 1994.

3. Bunting AH. Time, phenology and the yields of crops. Weather 1975; 30:312-325.

4. World Bank (Bos E, Vu, My T, Massiah E, Bulatao RA). World Population Projections 1994-5 Edition. Estimates and projections with related demographic statistics. Baltimore: Johns Hopkins University Press for the World Bank, 1994.

5. Food and Agriculture Organization of the United Nations/UN Fund for Population Activities/International Institute for Applied Systems Analysis (Higgins GM, Kassam AH, Naiken L, Fischer G, Shah MM). Potential population supporting capacities of the developing world. Rome: FAO, 1982.

6. Handler P. Invited comment. (Report of the) IIASA conference '76, 10-13 May. Laxenburg, Austria: International Institute for Applied Systems Analysis, 1976; 1:131-133.

7. Bunting AH, ed. Change in Agriculture. Proceedings of the first international seminar on change in agriculture, Reading 1968. London: Duckworth, 1970.

8. Hunter G, Bottrall A, Bunting AH, eds. Policy and Practice in Rural Development. Proceedings of the second international seminar on change in agriculture, Reading 1974. London: Croom Helm and Overseas Development Institute, 1976.
9. Bunting AH. Feeding the world in the future. In Spedding CRW, ed. Fream's Principles of Food and Agriculture. Oxford: Blackwell Scientific Publications, 1992:256-90.
10. Bunting AH. Look at it this way. Outlook on Agriculture 1991; 20:209-12.
11. Voelcker JA. Report on the improvement of Indian agriculture. Calcutta: Government of India, 1891. Cited in Randhawa MS. A history of the Indian Council of Agricultural Research. New Delhi: Indian Council of Agricultural Research, 1979.
12. Allan W. The African Husbandman. Edinburgh: Oliver and Boyd, 1965.

ISSUES AND PROBLEMS IN THE TRANSFER OF BIOTECHNOLOGY

David W. Altman

Technology transfer is a well-known concept where new technical information and knowledge is generated one place and then moves to another. Given the critical global challenge from population growth for adequate agricultural production systems, technology transfer assumes higher stakes. Increasing crop production, usually estimated to require a doubling of current levels,[1] does not seem feasible without a major contribution from new technology. While much of the developing world faces difficult conditions to meet growing demand for food, feed and fiber with limited resources, the term "transfer" could conjure up a variety of disturbing images—of an inept telephone receptionist switching an urgent call to random extensions with an indefinite time period on hold, or of canceled travel tickets with a series of rerouted connection slips that are redeemable at the whim of unfriendly agents behind the counter.

Biotechnology adds another dimension to the technical potpourri. Like secret ingredients for a special sauce, biotechnology can appear illusive and mysterious, while projecting a futuristic aroma that connotes higher value, an appealing flavor and thereby solutions for unappetizing problems. Little wonder, then, that a perception of controversy often occurs about biotechnology transfer, producing advocates and opponents over the prospect of letting the biotechnology genie out to roam free in farms and fields across vast ecological systems in the tropics and sub-tropical regions.

Departing from the intellectual battlefield, elusive claims of good or evil from plant biotechnology transfer to developing countries will be put aside. In addition, the intent of this chapter is not to evaluate specific practical uses of plant biotechnology because the contributions

Plant Biotechnology Transfer to Developing Countries,
edited by D.W. Altman and K.N. Watanabe. © 1995 R.G. Landes Company.

in this volume from authors in developing countries give a clear picture of real-life possibilities. Neither will the guiding principles of the many facilitating groups be elucidated, as representative organizations describe their outlook in a later section. Rather, I contend biotechnology applications require several key considerations, that are less consequential for agricultural development assistance with conventional technology transfer, and will discuss these points to put some other presentations in perspective.

By no means will such an examination avoid asserting a definite position. Transferring plant biotechnology is only useful if the activity serves to advance development. Any application of science depends on its effectiveness within the context of normal requirements for sensible development policy, as Professor Bunting points out in his chapter. Like new religious converts, I've witnessed the sudden "discovery" by many biotechnologists of the necessity in industrialized countries to use plant breeding, agronomy and extension in the aftermath of genetic engineering. Moving these applications to developing nations is an analogous shift in thinking required of the biotechnology practitioner, because outside of the developed countries different skills will be required to demonstrate that the potential products are profitable to producers and processors, acceptable to end-users and more attractive than traditional alternatives. The lessons from minimal commercialization experience with plant biotechnology in industrialized countries cannot be carbon-copied in less-developed nations.

Mimicking a typical development assistance project seems equally disastrous. The decade of the 1980s saw an adjusted average investment by donors of U.S. $10 billion annually for agricultural development.[2] Although per capita caloric consumption continued to go up,[1] several comprehensive comparative studies of bottom-line indicators reveal little or no progress. For example, IFAD[3] noted that there are more than 1 billion human beings living below the poverty line, over 80% are from rural areas where agriculture is the critical economic component, no progress has been made in reducing the overall percentage of the population affected by poverty over the last two decades, increases in poverty disproportionately affect women and the prognosis is for the number of people in poverty to increase, possibly to as many as 1.5 billion by the year 2000. If agricultural development was conducted as a business, and a hypothetical board of directors examined the $100 billion investment during the last decade in light of accomplishments, then many might question why all the managers haven't been summarily dismissed.

My objective is to highlight specific issues and problems for plant biotechnology transfer. Starting with a definition and an illustrative overview, four areas of particular concern are then presented to assist formulation of effective policy for management of biotechnology transfer. These subjects are: (1) access to capital, labor and support services;

(2) intellectual property rights; (3) regulatory issues; and (4) information and knowledge.

DEFINITION AND OVERVIEW

A reexamination of the definition of biotechnology could be useful for evaluation of near-term opportunities, which translate as probable successes to a skeptical world impatient for receiving tangible results after tantalizing promises. The Committee on Crop Terminology of the Crop Science Society of America has adopted the following definition of biotechnology:[4]

"Development of products by a biological process requiring engineering technologies, such as fermentation or controlled environments, or utilizing current technologies (such as recombinant DNA techniques) for the modification and improvement of biological systems."

Therefore, agricultural biotechnology would include some traditional and modern technology applications. Traditional biotechnology comprises applications such as biological pest control and food and beverage fermentation. Modern biotechnology includes new cell and tissue culture techniques, diagnostics such as monoclonal antibodies and nucleic acid probes and recombinant DNA technology. For agriculture, the high-visibility item, recombinant DNA technology, encompasses genetic engineering that incorporates transgenes into plants, animals and microorganisms, as well as certain products such as recombinant vaccines. The entire range of applications should be considered for separating near-term from longer-term opportunities in plant biotechnology transfer.

With this background, agricultural biotechnology could be reclassified according to probability for current implementation in developing countries. I would place the following technologies generally in the longer-term category: (1) aquatic biotechnology; (2) most strategies with transgenic plants and animals; (3) vaccines and most veterinary drug products; (4) environmental amelioration technologies; and (5) most value-added applications for food processing. Examples of more near-term opportunities would include: (1) fermentation technology; (2) specialized plantation commodity crops; (3) more established food-processing applications such as for beverages; (4) more established chemical and pharmaceutical manufacturing such as for diagnostics; and (5) micropropagation, particularly for forestry species and high-value crops.

Of significance is that this list of near-term applications for developing countries does not match usual assessments of high priorities for agricultural biotechnology from the point of view of markets in the industrialized countries. In practical scenarios, the neo-colonialist impression of handing out presents from a wealthy, distant relative does a disservice to the original goal. For many tropical crops, particularly plants producing high-value commodities that can bring export earnings for food imports, the cutting-edge expertise is found in developing countries. Currently, 42 developing countries depend on one or two commodities for over 70% of their export earnings, and two thirds of these countries are in Africa (see 1993 Annual Report of the Managing Director of the Common Fund for Commodities). However, commodity crops are often neglected for consideration of plant biotechnology transfer, particularly in planning activities with development assistance organizations.

Advanced applications, exemplified by completed field trials of transgenic plants, confirm that technology transfer has accounted for a small portion of global experimentation. Using an analysis which tends to over-emphasize participation by developing countries, Ahl Goy et al[5] concluded that less than 8% of transgenic field trials have occurred in developing nations based on a global database comprising information through the end of 1992. Of this number, there were virtually none that represented technology transfer which included formal agreements, joint development by donor and recipient partners and other elements of traditional transfer. Considering that the U.S. conducted over 1,600 individual field trials with transgenic plants in 1994 alone (D. Hatmaker, personal communication), the total of under 100 during 7 years in all developing countries indicates the improbable focus on this application for immediate utilization in most circumstances. Significantly, China has gone ahead with the first commercialization of transgenic crops, non-conventional virus resistance for tobacco, and has planted the largest area of a single transgenic crop, more than 1.5 square kilometers, by 1992.[6] From this perspective, Chinese agricultural biotechnology is the most advanced in the world (see chapter 10), but several issues such as intellectual property rights could impede broad utilization of this type of advanced technology without further international agreements for cooperation.

Foremost in putting common notions of plant biotechnology applications in the context of potential usefulness through technology transfer is the investment requirement. Substantial resources have been allocated to biotechnology research and development (R&D) and might benefit many other countries. The biotechnology industry in the U.S. alone invested $5.7 billion in 1992 for biotechnology R&D,[7] but little is known about private sector allocations specifically for technology transfer abroad. On the public side, the federal government of the U.S. had projected to spend about $4.3 billion in fiscal year 1994 for biotechnology R&D, but only $12.7 million or about 0.3% of available funds was for biotechnology designated to international agricultural development.[8] An earlier analysis for 1985 (ISAAA, unpublished data) indicated that biotechnology R&D in the U.S. represented about 55% of the global total, with Japan contributing almost 20%, and that agricultural biotechnology was nearly one fourth of the total. By any measurement, global investment costs are predominantly in the private sector, with the U.S.-based companies taking the lead, and appear extremely high for the few products brought to market so far, but tiny allocations are directed to applications suitable for developing countries.

LIMITATIONS FOR DEVELOPING COUNTRIES

ACCESS TO CAPITAL, LABOR AND SUPPORT SERVICES

Most science managers have underestimated the importance of access to capital, labor and support services. This observation has been valid for the business community as well as the academic community. The overselling of agricultural biotechnology by the private sector in the 1970s and 1980s demonstrates the pervasive attitude of minimizing these practical considerations as evidenced by early estimates, early estimates of time frames for commercialization of initial products by end of the last decade, and as recently as 5 years ago. Major corporations were predicting product sales of more than U.S. $100 billion by the end of the century, which is regarded as ludicrous by current analyses.

The reality is that most of these projects are complicated and require planning for unpredictable events, such as the regulatory environment, which can greatly increase expenses and which need unforeseen skills and services. Several anonymous examples from individual private companies can illustrate the size of some R&D programs to date for agricultural biotechnology commercialization: over $600 million has been spent for a growth hormone, over $35 million for biocontrol with an insect virus and over $30 million for one application of a transgenic crop. These figures are only for single companies, that either have not yet started product sales or are just commencing marketing, and do not reflect the total R&D of the industry for that product. Also, such cost guesstimates do not include basic research and related experimentation by public institutions, which were essential to advance the product development. Given these requirements for actual delivery of a product to farmers, the predominance of investment in R&D by the private sector should not seem very surprising.

Access to capital has become a very important problem for many of the transnational biotechnology companies as well as specialized biotechnology firms. The volatile stock markets of 1993 made new financing more difficult at a time when the drive towards commercialization has contributed to a rise in net burn rates and to a drop in survival indices.[7] Over the entire biotechnology industry in the U.S., 58% of companies are holding less than 2 years of operating cash, and the agricultural biotechnology companies' survival index fell by 5 months during 1992.

On the bright side, there are predictions of increasing availability of capital in many of the developing nations, particularly in East Asia.

The 1993 Survey of Third World Finance by *The Economist*[9] highlighted the increasing flow of private capital to developing countries, with foreign direct investment (FDI) accounting for 16.7% of gross long-term capital in 1992 vs. only 8.3% in 1981. Significantly, five Asian countries, China, Malaysia, Thailand, Indonesia and South Korea, attracted 35% of the global FDI for all developing nations (FDI, in $U.S. billion [percent of GDP] for China, Malaysia, Thailand, Indonesia, South Korea and all developing countries for 1991, respectively: 4.4[1.2]; 3.5[7.4]; 2.0[2.2]; 1.5[1.3]; 1.1[0.4]; and 35.9[1.1]). These statistics are impressive and underline the prevalent opinion among investors that East Asia has some of the best growth potential opportunities. Malaysia's FDI in 1991 of $3.5 billion is particularly remarkable because this figure represents 7.4% of the GDP and 20.5% of gross domestic investment. However, agriculture traditionally has been a poor competitor for these resources. Also, the indigenous agricultural biotechnology industry is relatively young and extremely small outside of the industrialized nations, so that local partners are not as readily available for joint ventures. Without careful planning there might be few possibilities to utilize this investment advantage for agricultural biotechnology.

Access to labor and support services is another limitation. Because of sluggish growth in the agricultural sector and industry restructuring, there is a tendency to assume that human resources are not a problem. Even during the lackluster year of 1992 in the U.S., the biotechnology industry increased the number of employees to almost 100,000 which was a 23% increase over the preceding year. While the job opportunities for scientists did decrease across the board, especially in academic institutions, the biotechnology disciplines saw a sharp increase in demand.[10] Crucial specialties in particular, such as intellectual property rights and regulatory affairs, have experienced shortages of qualified professionals as indicated by dramatic rises in compensation for managers in these fields.[11]

Another complication is the general trend for decreasing enrollment in many countries for both undergraduate and graduate students in agricultural sciences (e.g. Hoke[12]). These disciplines will be very important as commercialization of agricultural biotechnology products accelerates, and traditional agricultural-education programs will need to change rapidly to incorporate more training related to biotechnology skills.

Many agricultural biotechnology companies have undergone significant changes to cope with the increasing demands for specialized professionals and support services. Strategic partnerships are becoming more commonplace as the fad for vertical integration has generally waned.[7] Even large transnationals have looked for more alliances rather than keeping all operations in-house, e.g. Calgene/Rhône-Poulenc, Monsanto/Pioneer & Delta and Pine Land Co. and Danisco/Sandoz. In coping with evolving needs, many companies have spun off subsidiaries, such as Land-O-Lakes, Calgene and DNAP, or radically reorganized divisions that deal with biotechnology R&D, such as Ciba-Geigy, ICI, Monsanto, Shell, Kemira, Mitsubishi and many others.

Developing countries have the opportunity to learn from the biotechnology experience in the more advanced countries. If investment capital can be encouraged to prioritize agricultural biotechnology, then sufficient financial resources could be identified. Alternatively, planning for the general trend for decreased funding for agricultural development assistance from the traditional donor agencies[13] and tailoring programs for new opportunities such as biodiversity conservation and environmental protection could be prudent. In addition, training for the needed specialties must be given a high priority because this aspect of overall planning demands significant lead time. Competent plant breeders with good working knowledge of biotechnology can not be produced instantaneously, and the more obvious current shortages for experts in international patent law and regulatory affairs will not be solved quickly.

INTELLECTUAL PROPERTY RIGHTS

The global situation concerning intellectual property rights (IPR) is in a state of flux at the moment. Because current trends are unclear as a consequence of new developments, such as the proposed Patent Harmonization Treaty, the 1991 amendment to UPOV, the successful conclusion of the Uruguay Round of GATT negotiations, the implementation of the Biodiversity Treaty and other events, specific guidelines for IPR are not available. Various options for national IPR systems are possible, and as national capabilities for IPR-intensive areas like biotechnology increase, the tendency for harmonization with the global effort for stronger IPR will be more prevalent.[14]

One expert in this field, John Barton of Stanford University, has noted the growing extension of IPR into earlier and earlier phases of R&D.[15] For example, regular patent protection for plant varieties is becoming a common practice. While the validity of many of these patents on plant varieties is not definitive, the patents on genes for both animals and plants have received broad international acceptance among the industrialized countries. The National Institutes of Health (NIH) in the U.S. has even pushed the debate about gene patentability to include the possibility of protecting only portions of genes without specific, known functions. In Europe, patent applications have been filed based on receptor sequences rather than genes. Especially since the approval of GATT, evolving IPR with implications for agricultural biotechnology in industrialized nations will impact utilization throughout the world.

In the near-term, the UPOV Convention will probably have some of the most challenging limitations because many gene patents are still pending court decisions. Few developing countries to date have accepted UPOV guidelines,[16] although recent approval by countries such as Argentina signifies the potential for a substantial change. Greengrass[16] also noted that historical increases in investment from the private sector have been dependent on acceptance of UPOV protection and that GATT acceptance for Trade Related Aspects of Intellectual Property (TRIPS) could require more universal compliance with UPOV.

As plant biotechnology now has advanced with introduction of the first commercial products in the U.S. in 1994, plant variety protection will become an immediate concern. UPOV was revised in 1991, but these changes have not yet fully taken effect, and some observers believe developing countries can opt for compliance with the earlier UPOV accord. The following consequences of the new provisions from the 1991 agreement are an expansion of plant variety protection according to Greengrass:[16]

1. After certain transitional periods, all plant species will be protected rather than the current requirement of a minimum number of 24 within 8 years after accession to the Convention.

2. Breeder's rights will be extended to all production of propagating material. This will circumvent the use of tissue culture to avoid UPOV protection as well as largely curtailing the exclusion of farm-saved seed.

3. Harvested material also will be protected by breeder's rights. For example, cut flowers, which are exempted from protection under the original Convention, will now have coverage just like propagating material.

4. A principle of "essential derivation" is defined to extend protection offered for the "initial variety" when "(a) it is predominantly derived from the initial variety or from a variety that is itself predominantly derived from the initial variety while retaining the expression of the essential characteristics that result from the genotype or combination of genotypes of the initial variety; (b) it is clearly distinguishable from the initial variety; (c) except for the differences which result from the act of derivation, it conforms to the initial variety in the expression of the essential characteristics that result from the genotype or combination of genotypes of the initial variety." Exactly how this provision will be implemented is unclear, but the intent is to prevent simple selection from protected varieties as a method to avoid compliance.

5. There is now allowance for individual countries to protect plant varieties by granting other titles, particularly patents. This clause would be particularly useful for biotechnology applications where specific plant gene patents would be important, and the so-called "ban on double protection" with the original UPOV Convention would be eliminated.

Consequently, the general trend will be for more restrictions as a result of IPR on commercialization of plant varieties that could limit availability of agricultural biotechnology applications.

REGULATORY ISSUES

Agricultural biotechnology has given an impetus to increased, specialized regulations, usually broadly-defined as biosafety. While biosafety generally will be applicable to plants, animals, invertebrates, microorganisms, aquatic species and anything defined as a genetically modified organism (GMO), some products per se such as recombinant vaccines and bovine somatotropin can also be covered by these regulations. In practice, most of the experience to date for advanced applications has been with the pre-commercialization testing for crop species, so this discussion will be confined to transgenes in plants. The guidelines of the NIH for laboratory-based experimentation, or alternatively the 1986 guidelines of the OECD, could be referenced for the somewhat-standardized, international approach for laboratory practices which will be applicable in virtually all cases for agricultural biotechnology.

There is no binding set of international regulations similar to agreements on patents, plant variety protection or other IPR considerations. In 1991, UNIDO/UNEP/WHO/FAO concluded a voluntary code of conduct which has served as a model for some countries. Additional references can be obtained from ISNAR,[17] IICA (A report entitled "Guidelines for the Release into the Environment of Genetically Modified Organisms" was released in 1991) and CIIFAD.[18]

Less than 20 countries have adopted a formal set of guidelines for biosafety. Several of the developing countries' national guidelines that have been implemented are among the more restrictive systems in the world, which could be a partial explanation why only 8% of the global total for transgenic field experiments had taken place in these regions. Many nations have gone ahead and authorized field testing without the benefit of formal guidelines, given 28 countries had completed transgenic crop experiments according to Ahl Goy et al[5] through 1992 (At least three additional countries that conducted field trials were not included in this list). Seven of the 28 nations, or 25% of the total number, had not accepted plant variety protection according to UPOV.[16] China is one of these seven countries which will have considerable influence on any eventual international system because the Chinese releases are some of the most advanced in the world in terms of commercialization and distribution.

The important trend to consider is that regulations are increasing. Depending on how provisions of the Biodiversity Treaty are implemented for GMOs, these regulations could expand to require acceptance of certain universal guidelines and could even have ramifications for conventional genetic improvement. For the moment, no compulsory regulatory system for biosafety is required, but a proposed protocol from

UNEP's Expert Panel to be considered by the Conference of Parties in late 1994 calls for a "legally binding instrument" which will regulate internal, domestic releases.[19] Even so-called compromise approaches, such as a draft circulated by the Ministries of Environment of the U.K. and the Netherlands, imply that GMOs are intrinsically hazardous and require intensive, case-by-case review using a relatively restrictive delineation of risk assessment. The Chinese and American decisions to commercialize transgenic crops without mandatory labeling has drawn substantial criticism from the opponents of biotechnology, especially from some European countries, and these critics could agitate for trade sanctions as well as for efforts to impose some universal system. Conversely, the push to impose a universal protocol has resulted in some observers calling for possible withdrawal from the Biodiversity Treaty.[20] When conscientious critiques are heard from developing countries (e.g. Sihanya[21]), the importance of the lack of effective training for any type of biosafety system is emphasized.

On the other hand, there has been some movement to change several existing national guidelines in the direction of relaxing the regulations. The U.S. has begun "deregulation" since June 1993 of six crop species—corn, cotton, tobacco, tomato, soybeans, and potato—with extensive data from previous field trials. A 1993 editorial in the *European Biotechnology Information Service* raised the issue of deregulation for some species tested in the European community and cited the precedent of the U.S. action,[22] and the December 1993 White Paper of the European Union suggests the need for greater flexibility in the regulatory approach. Also, prominent scientists and various representative organizations have spoken out against overly-protective biosafety regulations (e.g. Miller et al;[23] Browning et al[24]).

For the moment, those interested in agricultural biotechnology should stay vigilant to monitor regulatory developments. Biosafety regulation will not disappear and in the most likely scenario will actually expand, but practical systems might be ultimately put into use if adequate information from competent agricultural experts is utilized by governments. By ignoring the activities of opponents that wish to block agricultural biotechnology applications, polarization of positions becomes the actual situation, and biosafety concerns could greatly disrupt available opportunities and/or, more significantly, create an enormous financial burden for actual implementation.

INFORMATION AND KNOWLEDGE

There is no adequate measure of information availability. However, the increase in workshops, meetings, publications and other sources of information that deal with agricultural biotechnology reflect the concern to keep informed. Moreover, because developing countries could be regarded as less advanced in many fields of biotechnology, the availability of sufficient, relevant knowledge can be a vital limitation. Simply put, lacking high-quality information means decision-making will suffer. The three factors noted above will all involve access to high-quality information as an important, underlying issue determining the severity of any potential problem.

Several sources of information about agricultural biotechnology are available at minimal cost. These include the following organizations that have started databases accessible through various media: (1) OECD, (2) the IBS of ISNAR, (3) the ICGEB under the auspices of UNIDO, and (4) SCREEN. IBS is publishing a series of reports that are designed to elaborate issues involved in biotechnology-research program management and policy formulation; three are widely distributed on biosafety,[17] on different governments' policy[25] and on national planning.[26] Specialized journals have also been established that are concentrating on biotechnology for a wider audience than just scientists, such as *Bio/Technology* which is published by Nature Publishing Co. and *AgBiotech News and Information* which is produced by CABI. The Dutch government has an initiative to sponsor subscriptions to the latter-mentioned journal, *AgBiotech News and Information*, for research institutes in developing countries. In addition, CABI has recently begun offering a CD-ROM database entitled PlantGeneCD that focuses on plant biotechnology. Also, some courses are available that are specifically designed to assist senior science managers from developing countries for agricultural biotechnology applications, such as through CIIFAD.

Newsletters could be a useful source of information, especially because several of the newsletters are emphasizing plant biotechnology transfer and most can be obtained without charge. One of the more established newsletters is sponsored by DGIS, *Biotechnology and Development Monitor*, which is published in both an English and French versions. USAID publishes *BioLink* as a quarterly newsletter from the ABSP project. Other examples of relevant interest are: CIAT's newsletter from the Cassava Biotechnology Network, *European Biotechnology Information Service Newsletter* of the Commission of the European Communities, *European BioNews* from SAGB and *SCREEN Newsletter*.

Information about the private sector is more difficult to obtain, especially concerning proprietary applications which involve virtually all private-sector R&D. Some details are provided in the sources noted above although those newsletters, journals, databases, classes, etc. are not exclusively dedicated to private-sector information. In the U.S. and Europe, there are some accessible reports, such as The Industry Annual Report distributed free-of-charge by Ernst & Young. The Office of Agricultural Biotechnology in the USDA provides information. In Europe, the Netherlands prepares regular information on the Dutch biotechnology industry which is also freely available, and other countries might be approachable for similar material. A European industrial organization, GIBiP, provides information as a service for interested individuals.

In addition, there are other sources of information if sufficient funding is available. Various bilateral and multilateral agencies will support projects that evaluate agricultural biotechnology opportunities. Private consultant firms that can be employed for short-term consultancies involving agricultural biotechnology are also proliferating. Outside professional services could be extremely valuable for consideration, particularly in assessing the patent situation for specific applications.

CONCLUDING REMARKS

Formal arrangements for biotechnology-transfer programs are diverse (For a review, see Altman[27]), and representative examples of each category are analyzed in Section III of this book. These programs are managed by (1) philanthropic foundations such as the Rockefeller Foundation and recently the McKnight Foundation, (2) DGIS, USAID-ABSP, UNDP, ADB, and other bilateral or multilateral agencies, (3) international agricultural research centers of the CGIAR, and (4) a few independent, new institutions like ISAAA, IIRSDA, ILTAB, and CAMBIA. A recent analysis by ISNAR of the CGIAR revealed that nearly 40 organizations throughout the world have a mission specifying plant biotechnology transfer (see chapter 19), which indicates the current increased activity in facilitating the transfer of this technology to the developing countries.

By perusing the contributions from the authors from developing countries, readers will note the importance of these formal initiatives, but also the wide array of activities that are the result of transfer in the broad sense of the word's meaning. Because any technology transfer is movement of information and knowledge from one location to another receptive place for a practical end, the flow of applications will happen as long as a need exists, even without a formal structure. The need for agricultural development is not in question, so biotechnology transfer will proceed. Can we foster an enabling environment for the transfer with careful contemplation and lessons from what already has been done?

From my perspective most of the commonly-identified opportunities are exaggerated and/or will not be accessible for near-term utilization. In many instances, traditional breeding or other conventional technologies might achieve the same results more efficiently. Without predicting eventual success, some examples could be cited where advanced technologies are only one option to be compared with conventional methodology: (1) virus resistance for tomato; (2) delayed fruit ripening for tomato and other crops; (3) transgenic maize for insect resistance; and (4) tropical-oil quality modifications. At the same time, for a variety of reasons some traditional methods can be inappropriately continued, and the actual costs and benefits of some conventional technologies, such as pesticides, frequently are not carefully evaluated.[28] Agricultural biotechnology alternatives should be given full consideration when they are available. Another caution is that many assessments of certain agricultural biotechnology applications are derived from data in industrial countries, and these same applications could be dramatically different in other regions of the world, such as the impact that ivermectin compounds donated by Merck have had in the control of river blindness in Africa.

Therefore, plant biotechnology transfer to developing countries holds out options that could turn away negative images, if special requirements are addressed and wise policy formulation occurs. The term "transfer" could just as easily relate to changing the legal title to a prized possession to include more members of the family.

ACRONYM ABBREVIATIONS

ADB	Asian Development Bank
ABSP	Agricultural Biotechnology for Sustainable Productivity
CAMBIA	Center for the Application of Molecular Biology in International Agriculture
CGIAR	Consultative Group for International Agricultural Research
CIAT	Centro de Investigación Agrícola Tropical
CIIFAD	Cornell International Institute for Food, Agriculture and Development
DGIS	Directorate–General for International Cooperation (The Netherlands)
FAO	Food and Agriculture Organization
GATT	General Agreement on Tariffs and Trade
GIBiP	Green Industry Biotechnology Platform
IBS	Intermediary Biotechnology Service
ICGEB	International Centre for Genetic Engineering and Biotechnology
IFAD	International Fund for Agricultural Development
IICA	Interamerican Institute for Cooperation on Agriculture
IIRSDA	Institut International de Recherche Scientifique pour le Développement en Afrique
ILTAB	International Laboratory of Tropical Agricultural Biotechnology
ISAAA	International Service for the Acquisition of Agri–Biotech Applications
ISNAR	International Service for National Agricultural Research
NIH	National Institutes of Health (U.S.)
OECD	Organization for Economic Cooperation and Development
SAGB	Senior Advisory Group on Biotechnology
SCREEN	Swift Community Risk Evaluation Effort Network
UNDP	United Nations Development Programme
UNIDO	United Nations Industrial Development Organization

UNEP United Nations Environment Programme
UPOV International Union for the Protection of New Varieties
     of Plants
USAID United States Agency for International Development
USDA United States Department of Agriculture
WHO   World Health Organization

REFERENCES

1. Waggoner PE. How Much Land Can Ten Billion People Spare for Nature? Task Force Report No. 121. Ames, USA: Council for Agricultural Science and Technology, 1994.
2. Anonymous. World aid for agriculture declining despite food crises. AgBiotech News and Information 1993; 5:379N-381N.
3. Jazairy I, Alamgir M, Panuccio T. The State of World Poverty, an Inquiry into its Causes and Consequences. New York, USA: New York University Press, 1992.
4. Anonymous. Glossary, Madison, USA: ASA-CSSA-SSSA, 1994.
5. Ahl Goy P, Chasseray E, Duesing J. Field trials of transgenic plants: an overview. Agro Food Industry Hi-tech 1994; 5:10-15.
6. Anonymous. China's transgenic trials. AgBiotech News and Information 1993; 5:86N-87N.
7. Burrill GS, Lee Jr KB. Biotech 94, Long-Term Value Short-Term Hurdles. San Francisco: Ernest & Young, 1994.
8. Anonymous. Biotechnology for the 21st Century: Realizing the Promise. Washington, USA: U.S. Government Printing Office, 1993.
9. Crook C. Third world finance. The Economist 1993; 25 Sept 93:44.
10. Spector B. Scientific career forecast for 1994 remains gloomy, as funding constraints and sluggish economy persist. The Scientist 1993; 15 Nov 93:1.
11. Silverman ER. Survey: biotech executives' salary hikes indicate industry's health. The Scientist 1993; 26 July 93:19.
12. Hoke F. Study sees alarming science undergrad dropout rate. The Scientist 1993; 25 Jan 93:1.
13. Anonymous. The State of Food and Agriculture 1992. Rome, Italy: FAO, 1992.
14. van Wijk J, Cohen JI, Komen J. Intellectual Property Rights for Agricultural Biotechnology: Options and Implications for Developing Countries. The Hague, Netherlands: ISNAR, 1993.
15. Barton JH. Introduction: intellectual property rights workshop. In: Baenziger PS, Kleese, RA, Barnes, RF, eds. Intellectual Property Rights: Protection of Plant Materials. Madison, USA: Crop Science Society of America, 1993:13-19.
16. Greengrass B. Non-U.S. protection procedures and practices---implications for U.S. innovators? In: Baenziger PS, Kleese RA, Barnes RF, eds. Intellectual Property Rights: Protection of Plant Materials. Madison, USA: Crop Science Society of America, 1993:41-59.

17. Persley GJ, Giddings LV, Juma C. Biosafety, the Safe Application of Biotechnology in Agriculture and the Environment. The Hague, Netherlands: ISNAR, 1992.
18. Lesser W, Maloney AP. Biosafety: a Report on Regulatory Approaches for the Deliberate Release of Genetically-Engineered Organisms—Issues and Options for Developing Countries. Ithaca, USA: CIIFAD, 1993.
19. Anonymous. Report of Panel IV. Nairobi, Kenya: UNEP, 1993.
20. Miller HI. The Biodiversity Treaty: A Bureaucratic Time-Bomb? Hoover Institute Essays in Public Policy. Stanford, USA: Stanford University, 1994.
21. Sihanya BM. Technology transfer, intellectual property rights and biosafety: strategies for implementing the convention on Biodiversity. AgBiotech News and Information 1994; 6:53N-60N.
22. Van der Meer A, Rohte O, Schreiber K, Connell J, Kioussi J, Lex M. Editorial: biotechnology regulation—experience leads to change. European Biotechnology Information Service Newsletter 1993; 3:33-34.
23. Miller HI, Huttner SL, Beachy R. Risk assessment experiments for "genetically modified" plants. Bio/Technology 1993; 11:1323-1324.
24. Browning CB, Baumgardt BR, Clarke NP, Davidson JM, Lacy WB, Lund DB, MacKenzie DR, Martin MA, Sander EG, Stern PE. Emerging Biotechnologies in Agriculture: Issues and Policies. Gainesville, USA: National Association of State Universities and Land-Grant Colleges, 1993.
25. Komen J, Persley G. Agricultural Biotechnology in Developing Countries; a Cross-Country Review. The Hague, Netherlands: ISNAR, 1993.
26. Cohen JI. Biotechnology Priorities, Planning, and Policies. The Hague, Netherlands: ISNAR, 1994.
27. Altman DW. Plant biotechnology transfer to developing countries. Current Opinion Biotechnology 1993; 4:177-179.
28. Pimentel D, Acquay H, Biltonen M, Rice P, Silva M, Nelson J, Lipner V, Giordano S, Horowitz A, D'Amore M. Environmental and economic costs of pesticide use. BioScience 1992; 42:750-760.

CHAPTER 3

ENVIRONMENTAL ISSUES FOR PLANT BIOTECHNOLOGY TRANSFER: A NORWEGIAN PERSPECTIVE

Anne Kathrine Hvoslef-Eide and Odd Arne Rognli

INTRODUCTION

Plant biotechnology in developing countries poses many of the same challenges as in the industrialized world, but also certain unique aspects have greater significance. First, many of the products of genetic engineering that can contribute towards a more sustainable agriculture, i.e. plant varieties with genetically-modified pest resistance will have greater impact on agricultural productivity in many developing countries than in most other parts of the world. This is because of the often high disease and pest pressures that occur in climatic regions of most of the developing countries. Second, few developing countries have implemented laws and regulations to handle the safe introduction of biotechnology products. Field tests of genetically modified plants (GMPs) in such countries could pose a threat to the local environment that has not necessarily been addressed during risk assessment of the same or similar GMPs in industrialized countries. Finally, most of the world's biodiversity and centers of genetic diversity for the most important crop plants can be found in developing countries. Both the present and future value of genetic diversity has to be addressed in assessing the risk associated with gene transfer from GMPs to wild relatives or wild populations of the same species.

Plant Biotechnology Transfer to Developing Countries,
edited by D.W. Altman and K.N. Watanabe. © 1995 R.G. Landes Company.

Norway is still in the very early stages of utilizing biotechnology in industry, and has primarily been concerned with building up competence within the fields of modern biotechnologies. When it comes to regulatory matters, Norway only recently (2 April 1993) put a framework law in place to handle contained use and deliberate releases of genetically modified organisms (GMOs). In many aspects we could compare Norway's situation with that of a developing country. We see the most advanced countries proceed with field trials and marketing of products while we are still implementing our regulations and addressing the biosafety issues for the first time in our own environment. Every natural environment in the world is unique in its own way, and we strongly resent an idea of biosafety that claims a product to be safe if extensively tested in only one or two environments of the world. In Norway the presence of nature is pronounced, which makes us conscious of nature's own integrity. There is a lot of natural wilderness compared to cultivated land, which makes the border between cultivated land and surrounding areas larger than in many other countries. Norway's situation makes us especially conscious of the potential risks developing countries could face if they uncritically adopt the results of field trials in another part of the world to be valid also for themselves. We therefore welcome this opportunity to address the issue of transfer of plant biotechnology to developing countries in light of biosafety.

THE POSSIBLE IMPACT OF BIOTECHNOLOGY IN DEVELOPING COUNTRIES

The significance of biotechnologies for the developing world can be illustrated by the example of Asian agriculture. In the early 1990s, Asia had more than 50% of the global population, over 70% of the world's farming families, but only 25% of the world's arable land.[1] At the beginning of the twenty-first century, the per capita land availability may be 0.1 hectare in China and 0.14 hectare in India.[1] Meanwhile the average annual Asian population-growth rate is 1.86%. The only way for countries like China and India to feed their growing populations is through continuous improvements in crop yields.[1] China has gone into large-scale exploitation of rice hybrids for this purpose. China has also had the most extensive field trials of GMOs of any developing country to date.[2]

In our part of the world, one often meets arguments that totally dismiss biotechnology as a potential tool to increase food production and thereby to improve the quality of life in many other areas of the world. The claim is that food shortage in some regions is a result of world trade and therefore purely a distribution problem, not one of production capacity. Others claim that increased production should not be a goal in modern agriculture at all, because overproduction and adverse environmental effects are major problems associated with high-input agriculture. We concede that the world is a backward place when it comes to distribution, with destruction of surplus food in the rich countries, while millions of people starve around the globe.

The example from Asia clearly shows that the developing world must be enabled to increase their productivity both in industry and agriculture in order to feed their own population and to have long-lasting, sustainable development. Among the greatest threats to biodiversity in the world are the increasing world population together with land-demanding cropping systems. World population is increasing in those parts of the world where the greatest biodiversity can be found. These regions also have the greatest potential for increased food production. In addition to deforestation of rainforests, the agricultural practice of burning forests to create new agricultural land, common in many parts of the poorest regions, is a great contributor to loss of habitats and biodiversity.

Waggoner[3] argues that increased food production through higher yields per unit land area is still the most important factor in sparing land for nature when the world population reaches 10 billion sometime in the not-too-distant future. Increasing yield levels without using more agricultural land will save land for nature, and thereby protect biodiversity. Biotechnology will probably provide the key for producing more food and other agricultural commodities from less land and water in the twenty-first century, without the adverse ecological implications associated with the expression of the full yield potential of high-yielding crop varieties through high-input agriculture.[4] Biotechnology may offer the small farmers utilizing landraces great benefits, as gene technology by intentional design may just alter the few crucial genes that confer, for example, disease resistance, potentially protecting the small farmers' crops far more effectively than any pesticide-spraying program, as well as being simpler to administer for the farmer. Furthermore, there is no need for building expensive infrastructures like those necessary to provide pesticides and fertilizers on farms. If this scenario is to become true, without the adverse ecological implications we see from modern agriculture today, the authorities in the developing countries need to be particularly careful about what types of biotechnology products they adopt. Biotechnology may offer environmentally-friendly technologies by providing alternative energy sources such as gasoline substitutes from crops or biogasses. Furthermore, biotechnology may reduce inorganic fertilizer reliance by replacement through genetic engineering of nitrogen fixation or symbiotic-microbe manipulation.

Developing countries not only need to build up know-how on the actual methods and techniques of biotechnology, but also on regulatory matters. There are several initiatives around the world today that are encouraging the developing countries to build a regulatory system suited for their needs. For overview of these, see Krattinger and Lesser.[5] If the developing countries acquire enough technology to perform genetic transformations themselves, they can carry out this work for less cost than companies in the industrialized world, address the most urgent issues in their own country and also use the local farmers' own adapted germplasm. Some biotechnology transfer initiatives already do this.[6] They broker agreements between companies that hold patents for the work in question and the developing country's own scientists. This enables each country's scientists to do the transformations that are desired/ needed in a particular crop for a particular country.[7] An important point to bear in mind is that the developing country must define its own needs. The companies/scientific institutions from the industrial parts of the world must only provide the technologies the developing countries really need and request for themselves.

Because most of our diet is dependent on plants, either directly or indirectly through domestic animals, we will concentrate this chapter on the environmental issues related to the use of *plant biotechnology*. Another reason to focus on plant biotechnology is the fact that most of the world's deliberate releases of GMOs have been with genetically modified plants (GMPs).[8]

PLANT BIOTECHNOLOGY TRANSFER TO DEVELOPING COUNTRIES

There are of course several techniques of biotechnology which may be of interest to developing countries. Even though this chapter deals with the environmental concern of plant biotechnology transfer, the different methods and their uses will be briefly discussed here, to point out where the environmental issues are most pronounced, either in a positive or negative direction.

MERISTEM CULTURE

Meristem culture is a way of cleansing vegetatively-propagated plants of internal bacterial diseases or viruses. The method is extremely valuable for making disease-free stock plants. Clean stock plants are in high demand, especially in climates where there are many virus transmitters, as in developing countries. The method does not, however, protect against reinfestation. It is, therefore, relatively expensive to keep healthy stocks of all the varieties most commonly in use. To be able to assure that the most valuable material is maintained disease-free, there is a need for a governmental program for large-scale meristem culture. For example, the necessity of governmental participation becomes clear when the situation in Norway is compared with small farmers using landraces in a developing country. Not even in a rich country like Norway will the farmers themselves pay the costs of obtaining disease-free stock plants. They rely upon a governmental program financing the cleansing of stock plants through meristem culture. If financed through proper channels and offered as a service to clean local landraces from diseases, it could be a positive contributor towards maintaining the biodiversity in landraces. It is, however, dependent on an active extension service that can make small farmers aware of the opportunity and the benefits. In addition, in order to maintain and not to decrease the biodiversity, careful consideration has to be taken that enough clones are cleaned each year and used as the basis for multiplication of local cultivars.

MICROPROPAGATION

Micropropagation is a method to overcome shortages of plant material or to produce masses of clonal material from elite stock (disease-free or genetically superior). This method could play a crucial role in producing enough plant material for the reforestation of rainforests, or the establishment of plantations of valuable plants instead of exploiting the rainforest. For example, growing rattan in plantations instead of collecting the plants in the rainforests will help preserve biodiversity. In addition to the harvest of rattan, the surrounding forest is often damaged to a large extent. In Ghana, where there is an increasing interest in production of manila furniture, The Ghanaian Forest Research Institute has calculated that the rattan in the rainforests will be near extinction in 4-years time at the present harvest rate (Grete Åmli, personal communication). There are many examples of shortage of seeds to maintain the rainforests at the present level, or even near this level. Micropropagation could, therefore, become an important tool for production of enough plant material. From an ecological point of view, it is also of vital importance that the number of clones is large enough to maintain a certain diversity.

PLANT BREEDING PURPOSES

There are a number of biotechnological methods used in plant breeding to increase variation and to improve efficiency of breeding methods, such as more cost-effective selection techniques, embryo rescue, somaclonal variation, in vitro mutagenesis, dihaploid production, protoplast fusion, recombinant DNA technology (genetic engineering) and in vitro selection. In principle, all these techniques are of interest to the developing world; lately most attention has been focused on genetic engineering. This focus is partly because of its powerful nature as a tool in plant breeding where natural barriers for moving genes between species no longer exist in principle. Gene technology opens up numerous possibilities for plant breeders, but also great concerns from environmental groups around the world. Of all the methods of biotechnology, gene technology (and protoplast fusion between species not capable of producing offspring by sexual reproduction) causes the most concern about ecological implications. It is not the technique per se that causes concern, but the potential for moving genes that may alter the organism's adaptation to the environment and invasiveness.

BIOSAFETY IN DEVELOPING COUNTRIES

Biosafety has been defined by Persley et al[9] as "the policies and procedures adopted to ensure the environmentally safe application of modern biotechnology." Most OECD countries have adopted the OECD guidelines on safety of gene technology, either into existing regulatory frameworks or into specially-designed gene-technology laws. Developing countries are to a large extent still lacking the regulatory framework to administrate gene technology, mainly due to lack of financial support and enforcement systems.[5] Notable exceptions are India, Mexico and the Philippines, which have established regulations and incorporated them in national laws. Argentina and Cuba have regulations in place, which have not been incorporated into laws, but established by ministerial decrees.[5] Finally, Bolivia, Brazil, Chile, China, Colombia, Costa Rica, Indonesia, Malaysia, Thailand and Zimbabwe either have ad hoc committees, or are in a more-or-less advanced stage of drafting regulations. There is a need to move faster, especially because field trials performed in developing countries exceed 70 trials already.

SYSTEMS OF RISK ASSESSMENTS AND REGULATION

The "case-by-case every case" philosophy is presently the basis for risk assessment and regulations of deliberate releases of GMPs in most of the industrialized countries. This philosophy has been very important in development of the OECD guidelines and the European Union (EU) directives. In the ongoing debate about regulations of biotechology, there are several calls both in Europe and in the U.S. for a simpler risk-assessment system, often called "product-based" or "science-based," as opposed to the "process-based" case-by-case system.[10,11] The reasons for these calls are mainly 2-fold. Firstly, results accumulated over recent years from a large number of field trials with many plant species show little or no evidence of unexpected ecological effects. One could therefore contend that, on a scientific basis, few arguments are in favor of excessively precautionary and costly risk-assessment procedures. Secondly, there are claims, especially from the biotech industry, that the present system imposes an unnecessary burden to the research which hinders the development and utilization of safe and valuable products. The U.S. has the most extensive experience with field tests of GMPs, and has introduced a simplified product-based procedure for field releases of six species, maize, soybean, potato, tobacco, tomato and cotton.[12] The EU is also introducing simplified procedures for certain plant species which have been thoroughly field-tested in Europe.[13] Traditional process-based regulations will be retained only for work involving pathogenic organisms and for deliberate releases of GMOs outside the low-to-negligible risk category.

The environmental issues related to biotechnology are in principle no different in the developing countries than in the industrialized countries. In cases where it is appropriate, experiences from field releases in other parts of the world could be used to perform a simplified risk assessment for releases in developing countries. Examples of such releases could be releases of well-known GMPs with familiar gene constructs and no risk of gene flow to the natural flora. However, developing countries are often the countries of origin (centers of genetic diversity) for our most important crop plants. In many countries local farmers are using landraces which are important sources of genetic diversity. Deliberate releases of GMPs in such centers of genetic diversity and among landraces require special attention. The results from field trials performed in the industrialized part of the world, under very different climatic and ecological conditions, cannot directly be transferred to other parts of the world. An impact analysis performed in a given geographic region for a crop plant that cannot hybridize with any species in the natural flora is not valid if the GMP is to be transferred to its center of origin.

In fact, if deliberate releases of GMPs are planned in the center of origin for a particular plant species, it is in the best interest of the whole world's population that such an application for release be particularly scrutinized and assessed. Centers of genetic diversity have been the most valuable resource for new germplasm for plant breeders to date. The introduction of gene technology in plant breeding is not going to change the dependence upon new germplasm for the continuous improvement of breeding populations. From these improved populations, valuable new superior lines can be developed to serve as the basis for genetic engineering in the future. If introduction of GMPs into such vulnerable environments is not carefully conducted, this recklessness could cause irreparable damage to those natural populations. Some might read this to mean that releases into those environments should never be performed. However, the crucial questions in relation to this concern are the gene construct that has been introduced into the plant, the extent of areas the transgenic varieties will cover compared to the natural populations of the plant, whether the transgene itself will confer a selective advantage to the natural population if spread and whether the introduced genes will be fixed in the natural populations or lost through genetic drift and/or natural selection.[14,15] Another possibility is to prevent genes from escaping into natural populations by introducing stable sterility in the GMPs.[15]

Miller et al[10] argue that over-regulating biotechnology causes exhorbitant extra costs to development of new products and acts actively as a deterrent for creativity and investments in biotechnology. Small and financially-weak biotech enterprises will obviously be at a disadvantage in an over-regulated and costly system. Over-regulation could also slow down the development and utilization of plant biotechnology in developing countries. Therefore, we will emphasize that we are not advocating a strict and bureaucratic regulatory system, but a flexible system that addresses the safety issues properly, and at the same time is promoting the development of the domestic biotech industry. As a conclusion it could be said that it may not be right to ask the developing countries to have the same level of *bureaucracy*, but it is vital to have the same level of *safety*.

The potential ecological risks following releases of GMPs fall principally in three classes: (i) direct effects of the transgenic genotype/cultivar, (ii) effects of the transfer of recombinant DNA (transgenes) to other populations/species/organisms, and (iii) indirect effects associated with the use of transgenic cultivars.

DIRECT EFFECTS OF THE TRANSGENIC GENOTYPE/CULTIVAR

These might be effects like increased weediness and invasion of natural habitats and production of secondary products (toxins, allergens) that might be harmful for humans, farm animals or other organisms. Although the potential for increased weediness is the risk most often highlighted, the experience from genetic engineering and field releases so far indicates that this risk is very small. This is true for crop species that have a long history of cultivation and where a high level of familiarity exists.[16] The problem of increased weediness might be different if genetic modifications are done with less-domesticated species such as grasses and forest trees and when the introduced genes confer changes in the characters that determine ecological tolerance and habitat preference.[17] So far, none of the field tests conducted have been of these types.

The problems with secondary products arising through genetic engineering pose, in principle, no other risk than that known from traditional breeding. However, because our experience of gene transfer with sequences from unrelated organisms is new, it is necessary to carry out thorough testing and risk assessment of each novel type of transgenic plant material. As our experience with field tests of specific combinations of gene constructs, plant species and environments increase, testing and risk assessment can probably be simplified. Methods for testing the presence of specific compounds and schemes for selective breeding for reduced contents have been developed for many crops. Similar procedures can also be used in the production of genetically modified cultivars to prevent undesirable genotypes being released into the environment and being used commercially.

EFFECTS FROM THE TRANSFER OF RECOMBINANT DNA (TRANSGENES) TO OTHER POPULATIONS/SPECIES/ORGANISMS

This is of two types, transfer of recombinant DNA to other species and/or populations of the same species via sexual hybridization (pollen flow) and nonsexual transmission of recombinant DNA, so-called horizontal gene flow. The existence of the latter possibility has so far not been established. Gene flow through hybridization may constitute a real ecological risk. Experience from monitoring of field trials with GMPs and from studies of hybridization among close relatives tells us that a transgene certainly will spread to neighboring close relatives or feral populations of the same species. There exists some possibility, even if it is small, for gene transfer. For most of the important crop species, this risk will in general be higher in the developing countries than, for example in Europe or North America because of the proximity to the centers of genetic diversity for many of these crops. Such gene transfers might constitute a threat towards the valuable diversity present in the centers of origin of plant species. The actual number of transgenes escaping is scale-dependent and will increase when transgenic cultivars are grown commercially. In cases where a transgene will escape, the risk assessment has to be based on the phenotypic effects of the gene in natural populations. We have to ask: "Can this gene do any harm if spread to wild populations?" One of the problems we are facing is the lack of ecological data to predict the fate of transgenes. Prediction might even be more difficult in developing countries due to more complex ecosystems. In general, transgenes that might result in adverse and unwanted ecological effects in natural populations should not escape from cultivation. Transgenic cultivars containing such genes should be constructed so that gene transfer is prohibited or virtually eliminated.

INDIRECT EFFECTS ASSOCIATED WITH THE USE OF TRANSGENIC CULTIVARS

The third type of ecological risk is indirect effects due to changes in cultivation practice following the use of transgenic cultivars. In the case of changes from highly-toxic herbicides to low-toxic and low-dose preparations, this effect may be positive in a country where herbicides are used regularly. To introduce herbicide-tolerant crops in rural areas in developing countries will necessitate the presence of that particular herbicide, which may not be beneficial to the small farmer. Utilization of plants expressing the Bt-endotoxin genes for the control of insect pests will have largely-positive indirect effects by reducing the use of highly-toxic insecticides. This might have a much larger impact in developing countries than in industrialized countries both due to the frequent use of insecticides in these countries, and the often improper use with associated health risks for the workers.

The three most popular modifications tested extensively in field trials so far are herbicide, virus and insect tolerance. The use of herbicides in conventional cropping systems is low in many developing countries. From an environmental standpoint, a large-scale use of herbicide-tolerant cultivars in developing countries would not be favorable because it will increase the general use of herbicides.

The most appropriate genetic modifications for developing countries would, in our opinion, be genes conferring tolerance against pests and diseases, and stress tolerance, e.g. salt tolerance. Very few field releases have been conducted with plants genetically engineered for stress tolerance. Due to the complex genetic background of many stress responses, such modifications are likely many years into the future. Genetically-engineered salt tolerance will confer increased competitiveness that might cause problems in species with weedy feral populations or a known record of invasiveness in natural habitats. Such modifications therefore need to be evaluated carefully before being released into the field.

Crop production in developing countries is very vulnerable to loss by insect damage which leads to frequent use of insecticides. The benefit of using insect-tolerant plants will be great in crops where a great amount of insecticides are being used, e.g. cotton. This crop alone contributes 10-15% of the total insecticide use in the world (D. Altman, personal communication). Transgenic plants expressing the Bt-endotoxin gene from *Bacillus thuringiensis* have been introduced in a number of species, e.g. tomato, tobacco, cotton, maize, potato and rice. Field trials confirm that the Bt genes give a high level of protection against insects. Many insect populations have developed resistance as a result of conventional use of Bt toxin as a biopesticide, and especially from the use of chemical insecticides. So far, only one example of resistance towards Bt is known, in field populations of diamond back moth (*Plutella xylostella*) from Hawaii and countries in Asia.[18] Following the large increase in the use of Bt as a conventional biopesticide and in addition the use of insect-tolerant plants expressing Bt genes, there is a great concern that field populations of insects will develop resistance against Bt. In view of the very positive impact the use of such transgenic plants might have in developing countries, it is of utmost importance to assure that the use of insect-tolerant transgenic cultivars in developing countries, as well as in the industrialized part of the world, is followed by proper regulations and regulatory oversight. Otherwise, the benefit could very soon be lost. A number of measures can be taken in order to reduce the risk of developing field resistance. Among the most effective will be the use of refuge areas, where insects are not exposed to Bt, and proper rotation systems.[18] It might be advisable to establish guidelines and maybe some form of regulatory oversight, for the implementation and use of transgenic cultivars in given cultivation systems.

NORWAY'S PERSPECTIVE ON GENE TECHNOLOGY AT HOME—AND ABROAD

The Norwegian Gene Technology Act with the regulations concerning deliberate releases of GMOs came into force on 1 September 1993. The importance of minimal risk for health and the environment, as well as the consideration of ethical issues concerning the use of the technology in each individual case, has been stated in the purpose of the law. The law also emphasizes the need to evaluate the benefit for society from the release into the environment of a particular GMO. The release should in addition preferably promote more sustainable development.[19] The Norwegian government has stated a positive attitude towards modern biotechnology and the possibilities such a technology may provide.

On the other hand, Norway is very much aware of the powerful tool such a technology represents, and the need to be able to refuse permission for a deliberate release if the release could lead to risks of an ecological nature or cause health problems. However, provisions for ethical consideration, benefits for society and promotion of sustainable development may be regarded as extra nets in which to "catch the fish" (i.e. refuse permission) in cases where it is not beneficial for society to pursue development of certain applications of biotechnology. In most cases it does not necessarily mean more difficulty to obtain permission in Norway than anywhere else. In those few cases where refusal cannot be based on ecological or health risks, authorities in other countries would probably have liked to have the legal basis for refusal of permission based on a total evaluation of the case, but have no law to support such a refusal and hence are forced to issue a permission. Krattinger and Lesser[5] claim that mixing biosafety and socio-economics within the same legislation would lead to bad science and poor economics and sociology, as well as delaying all introductions. We would argue that whether this is true or not depends on what stage in the evaluation process that the socio-economic dimension is taken into consideration.

At the basic research level, it is of course impossible to know ultimate impact from a particular project. Norway has solved this by stating in the preparatory text to the law that basic research is regarded as beneficial to the society per se. When a project approaches near-term applications or the stage of marketing, it is essential to give a product full and thorough evaluation. This evaluation should first of all be based on the risks for health and the environment. In addition questions about potential benefits of the product for the society, its contribution towards a more sustainable agriculture and whether the development and use of the product is in accordance with the prevailing ethical view in the society should be addressed.

Herbicide tolerance is an example where Norwegian authorities have the possibility to refuse certain cases, based on a case-by-case evaluation, but may give permission to others. It would be beneficial for the society if the agricultural practices were more environmentally safe, e.g. by using fewer pesticides. Herbicides have markedly-varied toxicological and eco-toxicological effects. In a case where introduction of a genetically-modified plant cultivar enabled the farmer to change from a more hazardous herbicide to a more environmentally-friendly herbicide, or to production with fewer sprays or reduced amounts of herbicide, it could be given a positive evaluation by the Norwegian authorities. The same herbicide tolerance, used in another crop where the agricultural practices would not lead to the same benefit, may be regarded quite differently. However, the potential impact on the environment from the spread of either the transgenic crop plant itself or the escape of transgenes via hybridization with wild relatives may be equal in both cases. The Ministry of Environment takes the final decision regarding issuance of permits for release, after having heard the arguments from different agencies and other ministries. According to the Norwegian Gene Technology Act, each case of deliberate release is evaluated separately, keeping the main goals mentioned earlier in focus.

Critics of the Norwegian Gene Technology Act have claimed that it is unnecessarily strict and that it will inhibit research and development in Norway. The framework law in itself may be looked upon as strict, but it is the implementation of the law which will determine whether Norway will handle these issues differently from other countries. Critics have also stated that the law may force researchers and production facilities to move abroad and thereby to escape the so-called strict Norwegian rules. That brings up another problem, and one which is vital to address in dealing with the developing world: what about transfer from the industrialized countries to the developing world of technologies that pose a risk?

TRANSFER OF PRODUCTS CONTAINING GMO OR PRODUCTION PLANTS TO DEVELOPING COUNTRIES

The Convention on Biological Diversity has raised this issue. Each party of the Convention shall make sure that the transfer of GMOs or production facilities where GMOs are handled is conducted in such a way that the receiving country is not put at risk. Norway has always supported the necessity for a binding protocol on biosafety rather than guidelines. Norway will continue to support the view that there is an obligation to report both export of GMOs and the intention of exporting GMO production facilities. Any GMO intended for export could be accompanied by a document informing the receiving country of the nature of the GMO in question and the environmental impact assessment. In this way the receiving country will be given the opportunity to refuse admission, if they wish to do so.

CONCLUSIONS

Every country has the right and the obligation to give an application for release of genetically modified organisms (GMO) a full evaluation based on the potential risks that a particular case poses for health and the environment, regardless of whether the same GMO has been evaluated in another country. It is especially important that developing countries are aware of the potential risks involved in releases near or in centers of origins of plants, as well as the presence of many landraces with valuable genetic variation. At the present time a case-by-case evaluation is still valid, even if extensive field trials have been performed elsewhere.

It is further delineated that a strict law and set of regulations does not necessarily mean stricter implementation when it comes to issuing release permits. Norway is taken as an example of a country with a seemingly-strict Gene Technology Act, that states the purpose of the Act is to ensure that the production and use of GMOs takes place in an ethically and socially justifiable way, in accordance with the principle of sustainable development and without detrimental effects on health and the environment. We believe a law that includes all aspects of public concern gives the public greater confidence and will contribute to a broader acceptance of the products when the time comes for entering the market. Implementation of the law will show whether such a law will delay all introductions; so far there have been four applications and four permits issued. The law will enable the authorities in Norway to stop the few cases that could jeopardize the safe application of biotechnologies.

REFERENCES

1. Sasson A, DaSilva EJ. Achievements, expectations and challenges. The UNESCO Courier, June 1994:11-15.
2. Altman DW. Plant biotechnology transfer to developing countries. Current Opinion in Biotechnology 1993; 4:177-79.
3. Waggoner PE. How much land can ten billion people spare for nature? Council for Agricultural Science and Technology (CAST), Report #121, Ames, Iowa, 1994:1-64.
4. Swaminathan MS. Biotechnology for beginners. The UNESCO Courier, June 1994:8-10.
5. Krattinger AF, Lesser WH. Biosafety—an environmental impact assessment tool—and the role of the Convention on Biological Diversity. In: Krattinger AF, McNeely JA, Lesser WH et al, eds. Widening Perspectives on Biodiversity, IUCN—The World Conservation Union and The International Academy of the Environment, Cambridge: The Burlington Press Ltd, 1994:353-66.
6. Altman DW. Technology transfer initiatives of the International Service for the Acquisition of Agri-Biotech Applications. AgBiotech News and Information 1994; 6:131-133.
7. Brenner C, Komen J. International initiatives in biotechnology for developing country agriculture: promises and problems. OECD Development Centre, Technical paper #100, OECD, Paris, 1994:1-60.
8. OECD—Field releases of transgenic plants, 1986-1992. An analysis, OECD, Paris, 1993:1-39.
9. Persley GJ, Giddings LV, Juma C. Biosafety—The safe application of biotechnology in agriculture and the environment. International Service for National Agricultural Research, The Hague, 1992.

10. Miller HI, Burris RH, Vidaver AK et al. Risk-based oversight of experiments in the environment. Science 1990; 250:490-493.

11. Ward, M. Do UK. regulations of GMOs hamper industry? Bio/Technology 1993; 11:1213.

12. Anon. USDA transgenic trials. AgBiotech News and Information 1993; 5(7):235N-236N.

13. Official Journal of the European Union, November 1994 (details to follow).

14. Raybould AF, Gray AJ. Genetically modified crops and their wild relatives—A UK perspective. Research report No. 1, Genetically Modified Organisms Research Programme, Department of the Environment, 1993:1-177.

15. Rognli OA. Ecological effects of releasing genetically modified crop plants into the environment (in Norwegian, with English summary). Report made for the Norwegian Directorate of Nature Management by the Department of Biotechnogical Sciences, Agricultural University of Norway, Ås, 1994:1-116.

16. OECD—Safety considerations for biotechnology: Scale-up of crop plants, OECD, Paris, 1993:1-40.

17. Regal PJ. Scientific principles for ecologically based risk assessment of transgenic organisms. Molecular Ecology 1994; 3:5-13.

18. Tabashnik BE. Evolution of resistance to *Bacillus thuringiensis*. Annu Rev Entomol 1994; 39:47-79.

19. Hvoslef-Eide AK. The Norwegian Gene Technology Act—the ethical perspective. In: Walin S, ed. Research Ethics Series No.5, CRE, Swedish Committee on Research Ethics (in press).

A Perspective on Socioeconomic Research on Plant Biotechnology Transfer for Developing Countries

David R. Lee

Although the applications of plant biotechnology research have scarcely begun to be realized, a considerable amount of attention has already been devoted by social science researchers to the potential problems of and prospects for biotechnological innovations. The range of technologies, issues and outlooks represented in the literature to date has been great, although much of this literature has, perhaps of necessity, been highly speculative, sometimes wildly so. Because of the early stage of plant biotechnology developments, a consensus view of the implications of biotechnology for social science research—similar to that which has developed, for example, with respect to the "Green Revolution"—has yet to develop.

This chapter is designed to provide a perspective on social science, and particularly economic, issues pertaining to plant biotechnology research in developing countries. The purpose here is not to catalog the numerous emerging technological innovations which may prove to be the subject of future research efforts, nor to provide a comprehensive review or critique of the current literature. Both of these types of assessments are available elsewhere. Rather it is to provide a perspective

Plant Biotechnology Transfer to Developing Countries,
edited by D.W. Altman and K.N. Watanabe. © 1995 R.G. Landes Company.

on selected emerging priority issues which can be expected to foster an increasing demand for social scientific research addressing the challenges of biotechnology applications in developing countries in the years ahead.

In what follows, the "biotechnology revolution" is first placed in the context of a longstanding history of continuous technological change in agriculture. In developing countries, the "Green Revolution" of the mid- to late twentieth century is particularly relevant to this discussion. This is true in part because the development of modern plant varieties (and the technologies which accompany them) can be seen as an intermediate step in the historical continuum between farmer varietal selection and subsequent advances in plant genetics and breeding, on the one hand, and genetic engineering on the other.[1] A number of common issues which have been studied and debated in the context of the Green Revolution—technology adoption and diffusion, structural change in agriculture, international trade effects and evaluating the returns to research—are reviewed for the insights they may contribute to our understanding of the likely future effects stemming from biotechnological innovations. Following this discussion, a number of emerging issues of particular relevance to biotechnology assessment are discussed; these include risk assessment, regulatory and intellectual property issues, the privatization of research and development,[1] and reassessing the role of the public sector in biotechnology research. Finally, ways in which developing countries can hope to achieve the greatest potential gains from biotechnological innovations are briefly identified and discussed.

TECHNOLOGICAL REVOLUTIONS IN AGRICULTURE

Revolutionary technological developments in agriculture are anything but new. The "biotechnology revolution" is only the most recent in a series of large-scale, transformative, productivity-enhancing developments in world agriculture, developments that, in most cases, began in developed countries and later were extended to the developing world. The mechanical revolution grew out of the Industrial Revolution and assumed its most striking form in the United States, where a succession of mechanical innovations in crop tillage, planting and harvesting transformed nineteenth century U.S. agriculture. In the twentieth century, a biological revolution began with the development of hybrid maize in the 1920s and later spread to plant breeding for many other crops, which in turn experienced substantial yield gains. A concurrent agrichemical revolution led to vastly increased usage of commercial fertilizers to stimulate crop yields and to the use of herbicides and pesticides to control weeds and insects. In addition to these developments, Cochrane[2] has argued that a fourth "managerial revolution" has characterized U.S. agriculture in the mid- to late-twentieth century, wherein farmers have made increasingly sophisticated use of management alternatives such as commodity futures markets, computerized management systems, and tax strategies as instruments for improving farm management and achieving higher returns and productivity.

In the developing world, the above developments have also had major, albeit uneven, effects due to the extremely diverse farming systems—ranging from subsistence to fully commercialized—characterizing agriculture in these countries. The dominant story of the second half of the twentieth century, in terms of its impact on food and agriculture in the developing world, has unquestionably been the diffusion of the "Green Revolution" package of improved plant varieties and complementary technologies. The use of plant-breeding techniques to develop modern high-yielding varieties of, most notably, rice, wheat and maize, combined with their greater responsiveness to irrigation, fertilizers and herbicides, has led to spectacular and extensively documented increases in crop yields in many parts of Asia, Africa and Latin America (see, for example, the 1988 CGIAR report by Anderson, Herdt and Scobie[3]). The overriding prominence of Green Revolution technologies in accounting for agricultural growth has been mirrored in socio-economic research as well. Whether of a supportive or of a critical nature, this research has responded over the past two decades to an agenda largely set by the perceived successes and limitations of the Green Revolution.

Because of the historical importance of technological change in generating agricultural growth, assessing the sources of productivity changes in agriculture has been a recurring theme in agricultural economics research for many years. The most widely shared view that has emerged, the "induced innovation" approach, is most commonly associated with Hayami and Ruttan.[4] Briefly, this approach hypothesizes that technological change endogenously responds to market price signals, specifically through farmers' attempts to save in the use of scarce (expensive) factors of production by substituting factors which are relatively more abundant (cheaper). Viewed in this manner, the farm mechanization revolution in the United States in the nineteenth century, for example, represented an attempt to save on costly labor inputs by substituting mechanical technologies which made relatively greater use of the abundant input, land, thereby increasing labor productivity.

More recently, the spread of modern varieties in Asia in the twentieth century has, under the induced innovation framework, been due in large part to technical changes that are "biased toward saving the increasingly scarce factor (land) and using the increasingly abundant factor (fertilizer) in the economy."[4] Hayami and Ruttan[4] describe how, for example, the growth in Asian rice yields in the twentieth century has closely (and inversely) followed declines in the fertilizer-rice price ratio, which, combined with the development of appropriate institutions and enhanced human capital, has enabled farmers to increase yields sharply. This framework thus views technological change as endogenously responding over the long run to the demands of farmers and the innovations of scientists, both of which in turn respond (directly or indirectly) to the relative price signals sent by the market.

The induced innovation framework is important to considering the potential impacts of biotechnological innovations because it can be expected that the adoption and diffusion of these technologies will follow many of the same patterns that have characterized the diffusion of other agricultural technologies, particularly modern varieties, in the developing world. The range of plant biotechnology innovations that are expected to be of relevance to developing countries is wide, including new diagnostic processes and tissue-culture techniques, the development of insect, viral, bacterial and fungal disease resistance, genetic engineering, genetic mapping and so forth.[5] However, the broad economic effects of these diverse technologies are likely to fall into a fairly distinct set of categories. These effects include: increasing yields, including crop productivity in marginal farming environments; lowering production costs; the development of new food products; the development of more environmentally benign technologies and procedures; enhancing input and food quality; and reducing dependence on nonrenewable energy sources.[6,7] This listing of potential effects illustrates not only the promise that biotechnology holds for widespread productivity improvements, but its significant potential for enhancing agricultural growth in marginal, as well as favored, production environments.

Similar to other "induced innovation" effects, the productivity of both land and labor inputs used in crop production should increase as farmers evaluate the marginal productivity and profitability gains from increased use of plant biotechnology innovations against investment in other inputs. One might predict, for example, that farmers will be able to achieve enhanced marginal productivity gains more cost-effectively through increased use of biotechnologically improved seed varieties versus more expensive land or labor inputs. The widely shared view that plant biotechnology will permit important qualitative improvements in agricultural inputs and outputs is less easy to consider within the induced innovation framework. This approach has indeed been criticized as limited in scope,[8] though it is perhaps the closest to a consensus paradigm with which to evaluate future developments in plant biotechnology.

FROM GREEN REVOLUTION TO BIOTECHNOLOGICAL REVOLUTION: SOME COMMON THEMES

The prominence of Green Revolution technologies in the debates over technological change and its effects in agriculture over the past three decades suggests that much of the future discussion surrounding the effects of biotechnology innovations will inevitably be conducted around many of the same issues. It is thus useful to review briefly several of the key issues that have been prominent in the recent past and how research on biotechnology impacts is likely to compare and contrast with foregoing work.

TECHNOLOGY ADOPTION AND DIFFUSION

A central focus of numerous studies of modern varieties has been the farm size and distributional aspects of technology adoption and diffusion. Modern varieties generate their greatest yield gains when accompanied by increased use of complementary off-farm inputs such as irrigation, fertilizer and pesticides, which may be expensive and/or unavailable to small or resource-poor farmers. Thus there are reasons to suspect a priori that resource-poor farmers may lag in their adoption of these improved technologies. Although often considered to be nominally scale-neutral, modern varieties and complementary technologies are typically more easily available to larger and/or more efficient farmers who have the capital to purchase off-farm inputs, for whom key inputs such as irrigation and credit may be more readily available, and who may enjoy easier access to physical infrastructure and the research/extension system. Numerous studies on the adoption of modern varieties and complementary inputs beginning in the 1970s confirmed that large farmers and those with access—to inputs, credit, infrastructure and information—were represented disproportionately among early adopters. More recently, the evidence has shown that smaller farmers have in many cases "caught up" in terms of adoption of modern varieties.[3,9,10]

It appears highly likely that the focus on farm size and distributional issues related to biotechnology adoption will, if anything, assume even greater prominence in the future. Policymakers, international donors, the public and researchers themselves will continue to focus attention on these effects. Consider, for example, the recent experience with the commercialization of bovine growth hormone (or bovine somatotropin, bST) in the United States. Despite the nominally scale-neutral nature of this technology and years of experiments that conclusively demonstrated its safety to animals and humans, public concerns over the effects of the adoption of this technology on small-farm profitability and the future structure of agriculture were the underlying factors largely responsible for the delay in governmental approval permitting the commercialization of bST and for various grassroots initiatives to delay its adoption through instituting labeling requirements and the like. While this may be an extreme case unlikely to be repeated in many developing countries that are more likely to be primarily concerned with food production, it suggests that concerns of this nature have become a permanent fixture on the public agenda.

Concerns over size, scale and distributional effects associated with biotechnology adoption are arguably inevitable. This is in large part due to the familiar "technology treadmill" effect. The early adopters of new agricultural technologies typically tend to be large and/or efficient producers who have access to and are able to afford off-farm inputs, who have access to information from researchers, extensionists and private sources, and who are able to manage the risks inherently associated with adopting a new technology. Early adopters are also those who tend to realize the greatest benefits from technology adoption, since they typically experience productivity increases before aggregate expansion in output begins to exert downward pressures on price. Unless government intervenes through price stabilization or subsidy programs, it is virtually inevitable that those who adopt late will be disadvantaged from the standpoint of basic market forces.

There are dimensions of biotechnological innovations which may alternatively moderate or exacerbate these effects. To the extent that these innovations are incorporated primarily in improved germplasm that will be acquired by farmers through improved seeds, the benefits can be relatively easily captured by farmers without significantly altering production and management practices, and their dissemination can build on existing plant-breeding efforts.[11] Similarly, new plant biotechnologies offer the potential to target farmers' needs due to what Walgate[12] has termed their "greater capacity to respond to precise user needs in matters of soil deficiency, drought, salinification, pests and diseases." Yet, whose needs will be prioritized? Trends in technology development and diffusion, including the increased privatization of biotechnology research and commercialization, suggest that the spread of biotechnological innovations will be highly correlated with the existing distributional pattern of infrastructure, information and marketing networks, particularly for agricultural inputs.

STRUCTURAL EFFECTS

In addition to early-adopting farmers, the primary beneficiaries of technological change are typically consumers who benefit from the reduced food prices and higher real incomes made possible by greater food availability resulting from higher farm yields. These consumers may be either urban consumers or rural farm and nonfarm households who are net food buyers. As Lipton and Longhurst[9] have emphasized in the case of modern variety adoption, this means that the primary concern with technology adoption from a poverty standpoint has to be with those farmers who do not adopt, for whatever reason. These farmers not only do not gain from the productivity-enhancing effects of technology, but suffer from the price-decreasing effects of technology diffusion in the aggregate. Moreover, due to the gains accruing to early adopters, late or nonadopters are placed at a further competitive disadvantage in realizing scale economies in marketing and in their ability to outbid competitors for land that may be used to enlarge their farms and permit yet further scale economies in production.

Nonadopters of new technologies are not located randomly, however. For many reasons—access to irrigation, land quality, availability of infrastructure, human capital variables (education, training, etc.) and so forth—differences arise in the extent to which farmers and communities as a whole experience the benefits of technological change. The Green Revolution provides ample evidence that, in the aggregate, these forces have led to what Lipton and Longhurst[9] call the "localization of regional disparity." More recent studies of "second generation" Green Revolution effects have given some evidence that these initial disparities have been mitigated and in some cases reversed by rural economic diversification, development of a dynamic off-farm sector and participation of the landless and other disadvantaged groups in rural industry and micro-enterprises.[13]

A related issue which has received considerable attention in the Green Revolution literature concerns the labor market effects of technology adoption. This is a highly complicated issue, with numerous offsetting effects. The increased use of off-farm inputs associated with modern varieties—inputs that in many cases are labor-displacing—tends to create a declining share for labor among total factor inputs, as Alauddin and Tisdell[10] document in the case of Bangladesh, for example. However, absolute levels of labor inputs may increase for a variety of reasons: greater incidence of multiple cropping, increased production levels and increased post-harvest throughput generating additional labor demands in storage, milling and transportation. Net returns to labor also depend crucially on the interaction of various factors including the mix of family to hired labor, the degree of access to land of hired labor and the real wage effects in individual localities.

The Green Revolution experience is relevant to plant biotechnology developments because many of the same issues likely will be raised and will require the demand of social science researchers. The issue of "appropriate biotechnology" already has received considerable attention from those concerned with biotechnology's potential adverse structural and distributional effects. There are several means to assure that these adverse effects will be moderated. First, adequate knowledge of low-input, noncapital-intensive farming systems can be generated such that appropriate biotechnology research can be identified which may address the needs of these farmers.[12,14] This may be challenging as indigenous farming systems are highly diverse and incorporate intercropping and multiple cropping practices, while many of the new biotechnology innovations can be expected to have a narrow genetic base.[15] Second, biotechnology research can be focused explicitly on crops and problems of importance to resource-poor farmers. Examples include improved nutrient availability, built-in pest and disease resistance, and potential improvements in local processing of agricultural raw materials.[14] Finally, the research/extension infrastructure that provides inputs and information to resource-poor farmers can focus explicit attention on the needs of these farmers; this includes institutions ranging from the International Agricultural Research Centers[16] to institutions providing local adaptive research and extension services.

As has been historically the case, however, large and/or efficient farmers with access to inputs, infrastructure and information can be expected to be at the leading edge with respect to the adoption and diffusion of biotechnology. In addition, some specific aspects of biotechnology innovations and the environment in which they are being introduced suggest that the impediments to assuring an equitable distribution of these technologies and their resultant benefits will be even greater than in the case of modern varieties. These factors include the often narrow genetic base of biotechnology innovations, the increasing privatization of biotechnology research and development, the significant physical and human capital requirements of cutting-edge biotechnology research and development, the ability to recoup R&D costs through pricing strategies built on patent protection, and the widespread decline in the growth of support for public-sector agricultural research, that, more than its private-sector counterpart, is more accountable to public needs. In view of these and other factors, Barker,[6] for example, has argued that it is "almost inevitable" that the gap in technological progress will widen between industrialized and developing countries on the one hand, and between technologically-advanced and less-advanced developing countries, on the other. If the limitations of the Green Revolution are not to be repeated, explicit attention must be devoted early on by research organizations, donors and government policy-makers to assuring that the benefits of the biotechnological revolution are as broad-based as possible.

INTERNATIONAL TRADE EFFECTS

International trade effects of plant biotechnology innovations promise to be both similar to and different from trade effects associated with previous technological developments. The 1989 OECD report on biotechnology impacts summarizes four potential trade-modifying effects of biotechnology; these include (1) trade creation through the exporting of entirely new products, and trade shifts stemming from (2) the creation of new substitutes, (3) the introduction of new production processes affecting factor proportions and comparative advantage and (4) the reduction of material inputs to production (the "dematerialization" of production). Like other technological innovations, the primary economic role of biotechnology development is to increase productivity and to decrease the costs of production. Thus there is a direct relationship between these innovations and international trade patterns built on comparative advantage principles. The trade impacts of biotechnological innovations are inevitable and likely will be significant.

Without question, the primary concern of developing countries in this area has been the issue of potential trade shifts due to the development of substitute goods and processes which may alter established trading patterns, particularly for products in which these countries currently enjoy comparative advantage. The example of the development of high-fructose corn syrup (HFCS) in the United States, Japan and Europe is often cited. In the U.S., consistent with the induced innovation model, high sugar prices in the 1970s, supported by highly protectionist U.S. farm policies which buffered U.S. sugar producers from international competition, provided the stimulus for the development of the corn-sweetener industry. The HFCS industry grew rapidly such that by 1981, only 67% of internationally-traded sugar originated from developing countries, down from 90% only 6 years previously.[17] By 1987, replacement of sugar by HFCS had reached 6 million tons.[18]

There are many other products for which biotechnological innovations in the recent past or near future hold the potential to displace developing country exports or shift current trading patterns. These include the development of other starch-based sweeteners, sweeteners based on the natural "super-sweets" such as thaumatin, numerous cocoa butter substitutes, vegetable oil substitutes, development of substitutes for vanilla and other flavorings, single-cell proteins for use as animal feed, a pyrethrum substitute and in vitro plant propagation and cloning. Many observers have noted that biotechnology offers the widespread potential for industrial-type production of agriculture's "end products"—oils, starches and proteins—such that the raw materials containing those end products increasingly become interchangeable.[15,17,18] To the extent that this occurs, agricultural trading patterns may become less a function of locational attributes (for example, tropical environments) and more a function of the traditional determinants of industrial comparative advantage, including research and development capacity, manufacturing plant efficiency and scale economies, and managerial expertise.

The implications of these developments for low-income countries are numerous and, in the extreme, portentous: further declines in the terms of trade due to increased global production capacity; reduced import markets for products in which importers have developed alternative sources of supply; loss of traditional export markets on which developing countries may rely for a significant share of export earnings, with accompanying macroeconomic implications; a loss of traditional sources of income and employment; and a severe blow to export-oriented growth strategies being followed by many developing countries. Some developing countries may be able to share significantly in the benefits accruing from biotechnological development; often these are considered to be larger and/or middle-income countries such as Brazil, India, Indonesia, Mexico, Thailand, etc. These countries share a number of important attributes including large domestic markets, a strong agricultural production and trading capacity, a broad human-capital base, and domestic biotechnology research and development capacity. In other cases, public policy changes may have a significant and as yet unforeseen impact on global trade outcomes. Consider, for example, the role of protectionist U.S. and Japanese sugar policies in leading food processors to search for cheaper sugar substitutes and the resultant effects on the global sugar trade.

ASSESSING THE RETURNS FROM TECHNOLOGY RESEARCH

Since the pioneering work of Griliches in the late 1950s evaluating the returns to research in hybrid corn, an extensive literature has developed which assesses the economic returns to investments in agricultural research and extension (see Echeverría[19] and Evenson[20] for recent reviews of this research). This research has been wide and varied, addressing returns to research in many crops, employing various analytical methodologies and representing both *ex post* and *ex ante* approaches. While much of this work has focused on agricultural research in industrialized countries, the experiences of developing countries, particularly in Asia and Latin America, also have received extensive attention.

Of the many lessons learned, certainly the most important is that the returns to agricultural research are almost universally estimated to be high, regardless of crop, location or analytical methodology employed. Of the 120+ studies reviewed by Echeverría,[19] for example, the estimated annual rates of return to research typically fall in the 30-50% range, and in only two cases are these rates of return estimated to be negative. Extensive treatment has been given to possible sources of bias in generating rate of return estimates—mis-specification of supply function shifts, spillover effects, public-private research interactions and so forth—but a strong consensus exists that the basic result holds widely. This is a primary reason behind the oft-stated policy prescription that the public sector invest more in agricultural research.

Despite this result, a number of factors strongly suggest that rates of return to investments in conventional technologies—irrigation, conventional plant breeding, fertilizer and chemical use, for example—will decline in the future. First, because of the discontinuous nature of scientific innovation, subsequent gains will be more expensive to generate in the laboratory and experiment station and will require more precise matching of technology development with crop- and location-specific needs.[21] Second, due to the logistic-type nature of technology adoption, those easiest to reach and in many cases those who, because of farm size and efficiency, are likely to generate the greatest productivity gains, already have been reached; subsequent diffusion of conventional technologies will have to focus on those more difficult to reach and who are producing in more marginal, less-productive areas. Third, no longer is it possible to focus agricultural research on productivity gains alone. Subsequent technology development, particularly that provided by the public sector and in industrial economies, will have to be compatible with broader social goals, including environmental preservation, the maintenance of genetic diversity, and increasingly-stringent regulatory constraints. Finally, publicly-funded research, which has accounted for much of the research on conventional agricultural technologies in many countries, is experiencing declining rates of growth globally.[22] Given the long lag between research investments and payoffs, the consequences of this decline may not become widely evident for some time, but it inevitably will reduce the rate of agricultural innovation and ultimately the growth of productivity.

What then are the implications of plant biotechnology research for estimating rates of return to agricultural research in the future? In terms of the key underlying question, the inherent productivity-enhancing and economic potential of plant biotechnology developments, it is far too early to tell whether the same consensus that has developed with regard to the effects of modern varieties and associated Green Revolution technologies will characterize plant biotechnology as well. Some observers have argued that due to both potential market incentives and technical factors, the earliest commercialized biotechnology innovations are likely to be among value-added products—horticultural products, industrial and export crops, and livestock and poultry products.[23,24] If this is the case, then developing country impacts should be highly selective in nature. If the initial evidence on recently commercialized biotechnology products such as bST and the bioengineered tomato is any indicator, then the returns to plant biotechnology innovations may be longer in coming and initially less powerful than earlier predicted.

A second question is whether the assessment of economic returns from technology research itself will receive the attention it has in the past. With public funding, the returns to research are a legitimate public policy question. Demands for this type of information by governments, donors, international agencies and other publicly-funded entities have been the primary motivating force behind the extensive past research in this area. However, with the greatest share of biotechnology research occurring in the private sector, there will be relatively less demand for public information on returns from research to justify public funding; when demand for this information does exist, it may stem from private sources and may, for competitive reasons, remain proprietary. This trend may be substantially mitigated in developing countries, where much biotechnology research is likely to remain in the public sector. Thus the returns to publicly-funded biotechnology research in developing countries and the payoffs from biotechnology transfer to developing countries will remain important public policy issues.

A final and opposing trend which may reinforce and reshape the public's demand for information on biotechnology's impacts will be the above-mentioned broader context within which technology development and diffusion must now be considered. While many of the benefits of plant biotechnology innovations may be expected to accrue to traditional private-sector beneficiaries—industrial firms, early-adopting farmers, food processors, etc.—in order to realize those benefits, various public policy concerns will have to be allayed in the context of authorizing research, field trials, commercialization and public consumption. Although this is likely to be a particular concern in industrialized countries, it also will be of concern in developing countries as well. These forces likely will continue to stimulate the demand for research on economic and related impacts of plant biotechnology research, particularly in order to alleviate public concerns regarding public health and food safety.

EMERGING ISSUES IN SOCIAL SCIENCE RESEARCH ON PLANT BIOTECHNOLOGY

A consideration of how traditional areas of social science concern may be redirected in light of developments in biotechnology research leads naturally to the question of what new and different issues will become highlighted in social science research on plant biotechnology. This section briefly discusses some of these emerging issues.

RISK PERCEPTIONS AND ASSESSMENT

Consumers, particularly in developed countries, are becoming more concerned about food safety issues in general. Those food products which make use of biotechnological innovations in altering products designed for direct consumption are likely to become targets of these concerns, in much the same way as have food irradiation, pesticide residues, animal antibiotic and hormone treatments, and other perceived (by some) public health and food safety issues. The recent public debates surrounding the use of bST and the bioengineered tomato suggest the confrontative nature of likely future debates. Bioengineered industrial products that are one or more steps removed from direct human consumption, including those used in food processing, are less likely to be subject to these concerns.

One implication of the importance of risk perceptions is the need for social science research that addresses these concerns, their determinants and consequences. For example, a recent study by Florkowski et al[25] in the United States illustrates the importance of these issues. A national sample of over 1,100 households was surveyed regarding their attitudes toward bioengineered porcine somatotropin (pST) and projected changes in eating behavior as a result of the availability of pork produced using pST. Nearly one-half of the sample (47%) responded that they were "concerned with the use of [any] recombinant DNA technology to produce a product for human food." One-third of the households surveyed suggested that they would consume less pork as a result of pST use. When advised of the benefits of pork produced using pST—the availability of a cheaper and leaner product—56% and 54% of respondents, respectively, indicated that they would not eat more pork in spite of these benefits. Differences in age, gender, education and location were found to influence consumers' perceptions and predicted behavior.

Surveys of European consumers' attitudes toward biotechnology and generic engineering have been conducted in 1991 and 1993 by the EC Commission[26] and are also illustrative. Factors found to influence consumer attitudes toward biotechnology included a wide variety of factors: consumers' philosophies, values and ethics; their understanding of objective information pertaining to biotechnology and genetic engineering; their trust (or distrust) of specific information sources; nationality; and a number of socio-demographic variables. Consumers were found to be less optimistic about the potential contributions of biotechnology to social welfare than any other technology identified (telecommunications, information technologies, etc.), except for space exploration. At the same time, the 1993 survey revealed a very high correlation between level of "optimism" and objective knowledge about biotechnology and genetic engineering. Sources of information also were found to differ widely in their perceived reliability, with environmental and consumer organizations and universities found to provide the most dependable information. Although risk perceptions varied widely across countries, the public's demand for governmental controls on biotechnological applications was overwhelming. These results suggest several conclusions: first, the importance of consumer perceptions regarding biotechnology, particularly because those same consumer perceptions influence the regulatory environment; second, the critical role of information in influencing consumers' knowledge and risk perceptions (and, in turn, attitudes); and third, the importance of providing information that is perceived as reliable, objective and not influenced by a particular viewpoint being promoted.

Although it could be argued that these two examples reflect products or circumstances which are nonrepresentative of consumers' perceptions in the developing world, the finding of significant consumer concerns regarding food safety and risk perceptions is a generalizable one. Many consumers have a general, though often ill-defined and difficult to articulate, set of concerns that relate to processes and products which defy easy or intuitive understanding. Given the sophisticated processes underlying plant biotechnology innovations, consumers' risk perceptions must be kept in mind. Moreover, if the experience in the U.S. is any indication, the definition of "risk" itself and associated guidelines and assessment methodologies may often become public policy issues. To the extent that plant biotechnology developments—for example, improved seeds, industrially-produced inputs, etc.—are removed from direct human consumption, there may be less concern. There also may be substantially less concern in developing countries geared toward the production of adequate food availability and relatively less concern about consumer perceptions of food safety.

LEGAL, REGULATORY AND INTELLECTUAL PROPERTY ISSUES

Public policy concerns with institutional, legal and regulatory issues regarding biotechnology transfer will assume much greater prominence than has been the case with Green Revolution technologies. This prominence stems from several sources. First, to the extent that food safety and public health concerns are relevant, appropriate regulatory authorities (counterparts of the U.S. Food and Drug Administration) will become involved. Second, concerns over the consequences of the release of genetically engineered organisms into the environment through field testing and commercial release may prompt the involvement of environmental regulatory bodies. Finally, the dominant role of the private sector in the development and commercialization of plant biotechnology innovations may spark renewed concern over industrial structure, research and development policies, and the pricing strategies of private firms which are involved in food production. Already, the issues of concentration and consolidation in the biotechnology industry as a result of the acquisition of many smaller firms by the large multinational chemical and agribusiness firms have been raised as public policy issues (see, for example, Hobbelink[15]). Given the stricter regulatory environment and greater enforcement capacity generally existing in industrialized countries, it remains to be seen to what extent these concerns will be translated to the developing world.

One area which will remain of intense interest to developing countries, however, concerns the patenting of biotechnology innovations and intellectual property protection. A variety of methods can be used to protect biotechnology innovations including seed and breed certificates, plant patent and variety protection, invention and industrial design patents, copyrights and trade secrecy.[27] In addition to laws at the national level, a number of international agreements have created a framework for intellectual property protection, including plant variety protection. These agreements began with the Paris Convention of 1883 (as amended most recently in 1967) and include the Budapest Treaty on Microorganisms (1977), the Patent Cooperation Treaty (1978), the International Convention for the Protection of New Varieties of Plants (as amended in 1991) and, most recently, the Uruguay Round of the GATT negotiations. However, few developing countries are signatories to most of these treaties, and despite the tendency toward greater protection of intellectual property, including biotechnology innovations, the consensus behind this trend is anything but unanimous.

Evenson and Putnam[28] identify a number of distinct country perspectives on intellectual property and patent protection that depend in large part on whether a country is a exporter or importer of technology. Industrialized countries, such the United States, that export technology (or products embodying advanced technologies) and that have a significant indigenous inventive capacity, tend to seek strong patent rights as a means of protecting these industries in which they have a comparative advantage. Technology-importing countries that have a local-adaptive or "copying" capacity have little incentive to protect foreign technologies; rather, the incentives may be to "pirate" the technology and to use it as the basis for a local adaptive industry. Such countries may negotiate patent protection in return for opening up their technology markets. For countries that import technologies and have little or no adaptive capacity, strong intellectual property rights may lower the cost of technology transfer.

The legal and regulatory environment for plant variety protection is critically important in creating incentives for local biotechnology investment and R&D, and in ultimately influencing the incidence of benefits versus costs from biotechnology development. As such, the assessment of these relative gains and the extent to which they are governed by the legal and regulatory environment will be an important subject for future attention by researchers. As this area inherently involves the legal status of living organisms, it can also be expected to remain a highly contentious one.

PRIVATIZATION OF RESEARCH AND DEVELOPMENT

Few other biotechnology issues have received as much speculative attention by social scientists as the increasing privatization of research, knowledge and product development. There are several dimensions to this issue which have led to these concerns. First, plant biotechnology research is expensive due to the sophisticated nature of the technologies involved, the requirements for extended field testing, the bureaucratic and regulatory hurdles involved, and so forth. One estimate of global biotechnology R&D spending in 1990 puts the figure at $ U.S. 11 billion, of which two-thirds to three-quarters came from the private sector. The magnitudes of these capital needs in large part account for the attention that has been given to the financing of biotechnology research. The Office of Technology Assessment's (OTA) 1991 report on global biotechnology, for example, maintains that the reason for U.S. dominance in commercial biotechnology has been not only the quality of scientific innovation but the available pool of venture capital (as well as federal funding of biotechnology research).

Second, the biotechnology industry itself is widely viewed as a "leading edge" industry that will become a source of employment and economic growth in the next century. The often-cited OTA and OECD biotechnology reports, for example, are strongly cast in terms of enhancing the competitiveness of the biotechnology industries in the U.S. and other industrialized countries. How to assure most effectively the success and viability of this industry has become an important industrial-policy issue in countries such as the U.S. and Japan and in Europe. In large part, this accounts for the public policy emphasis put on plant breeders rights and the securing of intellectual property rights as a key mechanism for undergirding biotechnology investments.

With many of the innovations in biotechnology coming from the private sector, considerable anxiety has been expressed by a wide range of observers regarding the implications for biotechnology transfer as a result of private profit-seeking companies setting the priorities for diffusion of technologies which could help solve key global challenges such as adequate food production. This is particularly true in developing countries, where the human needs are relatively great and where biotechnology has been viewed by some as a "magic bullet" which will enable nations to surmount other critical obstacles to food security, including population growth and environmental degradation. In such an environment, there is widespread concern over private companies seeking to enhance profits through pricing and product availability strategies which may limit availability and/or increase the prices of biotechnology products. At the other extreme is the view that private companies are largely responsible for initially incurring the risk, investing the capital and generating the technologies to begin with, and thus should be given an opportunity to receive a fair return on prior investments as well as be provided sufficient incentives for future investments. This debate is further fueled by the fact that private-sector biotechnology research is highly concentrated in the industrialized countries. The obstacles to successful technology transfer to developing countries are many, including the availability of investment capital, the adequacy of scientific and technological infrastructure, legal obstacles to information transfer and the cost of biotechnology R&D.[29]

In this emerging environment, with the private sector assuming a greater role in the setting of research priorities (and consequently forthcoming technologies), Busch et al[8] have argued that there must be greater private-sector accountability. Attempts to monopolize domestic markets or to engage in anti-competitive pricing strategies may be addressed by individual governments through antitrust enforcement and other mechanisms. In the international environment, however, there are fewer mechanisms to guarantee accountability, and the appropriate role of the private sector, particularly in those countries lacking a strong traditional private-sector orientation, will remain an important public policy issue.

Whither the Public Sector?

In the United States, numerous other industrialized countries and in some of the larger developing countries, the public sector has played a key role in support of biotechnology development. For example, in the U.S, this has included mechanisms such as: strategic targeting of biotechnology development for economic growth; federal funding for biotechnology research; regulatory development; intellectual property protection; enhancing the development of university-industry relationships; and favorable reforms in the tax code.[7] In developing countries such as Brazil, India, Thailand and India, governments have supported biotechnology development through such measures as the development of institutes and laboratories for biotechnology research.[30] The public sector can be expected to be the dominant source, and in many cases, the sole source of funding for biotechnology R&D and transfer in most developing countries for the foreseeable future.

Despite this key public-sector role, the role of public institutions, particularly universities, in biotechnology development and transfer is undergoing change and increasing scrutiny. There are several reasons for this. First, although public-sector research universities in the United States played a key role early in biotechnology development and continue to play an important role, in absolute terms this investment is dominated by the private sector. In developing countries, however, public-sector institutions can be expected to continue to assume a central role in adaptive biotechnology research and technology dissemination.

Second, given the history of declining growth in agricultural research funded by the public sector across a range of countries, much of the research accomplished by scientists employed in the public sector will be jointly funded by and conducted in collaboration with private-sector firms. Such public-private partnerships may generate some significant benefits in terms of increasing biotechnology funding, orienting biotechnology research toward commercializable technology development and constructing "critical masses" of researchers addressing key technical issues. However, as numerous observers have noted, there are also costs associated with these partnerships, including the increasingly proprietary nature of research results, and delays and obstacles in the sharing of research results with the broader scientific community and the public. The latter clearly includes developing-country scientists critically dependent on basic science results in the developed countries. The public sector's role in setting the biotechnology research agenda will thus continue to be more limited than it traditionally has been in agricultural research.

Finally, to the extent that public-sector biotechnology research focuses on products that are associated with contentious public-policy issues—such as alleged public health and food-safety effects of bioengineered foods or products that in the future actually generate structural changes in the agricultural and rural economy, the public sector may open itself to criticism from many quarters. The case of publicly-funded research on the labor-displacing mechanical tomato harvester in California is often cited in this regard. If developments such as bST use lead to the structural changes in the dairy sector anticipated by some observers, a similar reaction can be expected. Public-sector institutions that generate products which lead to structural changes in developing countries, or public sector-funded institutions (research institutes, universities, the International Agriculture Research Centers, etc.) that are associated with these structural changes, can expect to encounter similar criticisms.

If anything has been learned from past technological revolutions in agriculture, including the Green Revolution, it is that structural changes must be anticipated at the outset of the research, development, commercialization and diffusion process. Social science researchers have a key role to play in analyzing the likely aggregate effects of technology development on economic variables such as prices, output markets and labor markets, and on the distributional effects of technology on different groups of farmers, consumers and the labor force. The social science tools are in place to be able to conduct *ex ante* analysis of biotechnology developments on these groups and on key market variables. Consider, for example, the recent *ex ante* work on the aggregate effects of bST on U.S. dairy markets by Kaiser and Tauer[31] and McGuckin and Ghosh[32] or on the cotton industry by Chiou, Chen and Capps.[33] It will be important to extend this type of analysis to developing countries that may experience significant distributional or market effects from biotechnology adoption.

IMPLICATIONS FOR BIOTECHNOLOGY TRANSFER TO DEVELOPING COUNTRIES

The many issues discussed above—whether stemming from traditional areas of concern or emerging questions uniquely associated with biotechnology developments—suggest that significant challenges will exist in assuring the broad-based transfer and diffusion of biotechnological innovations in developing countries.[34] As with Green Revolution technologies, the technical and economic obstacles to an efficient and equitable process of technology transfer, adaptation, innovation and diffusion will be great. Perhaps more importantly, however, the many institutional factors reviewed above—changing private- versus public-sector roles, intellectual property rights and patent protection, effectively dealing with food safety, environmental and regulatory concerns, and so forth—will create further challenges. These factors should prove particularly difficult to address.

However, a number of actions and interventions on the part of international organizations and developing countries themselves can facilitate the process of broad-based technology transfer. These include:

- Targeting of "appropriate" biotechnologies—that is, those technologies that meet specific developing country needs and constraints—by developing country governments, institutes and laboratories;
- Financial support for appropriate biotechnology R&D and transfer from bilateral and multilateral donors;
- Development of biosafety regulations for field testing and commercialization of biotechnology products;
- Training of developing country scientists, along with other mechanisms such as research networks, collaborative North/South and South/South research and training programs, etc.;

- Support for the work of the International Agricultural Research Centers and their collaborative work with national agricultural research systems;
- Development of legal and regulatory frameworks for intellectual property protection and plant breeders' rights which can create appropriate incentives for the transfer and adaptation of new technologies;
- Institutional changes and infrastructure improvements which will assure the availability of improved seeds and other biotechnology products to a broad-based population of farmers; and
- *Ex ante* analysis of relevant plant-biotechnology impacts.

As promising as they are, biotechnology innovations in and of themselves cannot be expected to solve the food security problems of developing countries. In fact, as has been reviewed above, biotechnological changes may well exacerbate as well as help solve some of the chronic problems facing these countries. The types of actions and commitments identified above will expedite, indeed they will be essential to, the equitable generation of benefits stemming from the many promising developments in plant biotechnology.

ACKNOWLEDGMENT

The author wishes to thank David Altman, Randy Barker and William Lesser for helpful comments on an earlier draft.

REFERENCES

1. Persley GJ. Beyond Mendel's Garden: Biotechnology in the Service of World Agriculture. Wallingford, UK: CAB International, 1990.
2. Cochrane WW. The Development of American Agriculture: A Historical Analysis. Minneapolis: University of Minnesota Press, 1979.
3. Anderson JR, Herdt RW, Scobie GM. Science and Food: The CGIAR and Its Partners. Washington, DC: The World Bank, 1988.
4. Hayami Y, Ruttan VW. Agricultural Development: An International Perspective. Baltimore: The Johns Hopkins University Press, 1971 & 1985.
5. Dart PJ. Plant production: introduction. Ch. 3. In: Persley GJ, ed. Agricultural Biotechnology: Opportunities for International Development. Wallingford, UK: CAB International, 1990.
6. Barker R. Socio-economic impact. Ch. 22. In: Persley GJ, ed. Agricultural Biotechnology: Opportunities for International Development. Wallingford, UK: CAB International, 1990.
7. Office of Technology Assessment, U.S. Congress. Biotechnology in a Global Economy. Washington, DC: U.S. Government Printing Office, 1991.
8. Busch L, Lacy WB, Burkhardt J, Lacy LR. Plants, Power, and Profit: Social, Economic, and Ethical Consequences of the New Biotechnologies. Cambridge, MA: Blackwell Publishers, 1991.
9. Lipton M, Longhurst R. Modern Varieties, International Agricultural

Research, and the Poor. Washington, DC: The World Bank, 1985.

10. Alauddin M, Tisdell C. The "Green Revolution" and Economic Development: The Process and Its Impact in Bangladesh. New York: St. Martin's Press, 1991.

11. Brenner C. Biotechnology and Developing Country Agriculture: The Case of Maize. Paris: Organization for Economic Cooperation and Development, 1991.

12. Walgate R. Making Biotechnology Appropriate—and Environmentally Sound. Ch. 12. In: DaSilva EJ, Ratledge C, Sasson A, eds. Biotechnology: Economic and Social Aspects: Issues for Developing Countries. Cambridge, UK: Cambridge University Press, 1992.

13. Sharma R. The New Economics of the Green Revolution: A Study of Income and Employment Diffusion in Rural Uttar Pradesh, India. Ph.D. Dissertation, Dept. of Agricultural Economics, Cornell University, 1990.

14. Bunders JFG, ed. Biotechnology for Small-Scale Farmers in Developing Countries: Analysis and Assessment Procedures. Amsterdam: VU University Press, 1990.

15. Hobbelink H. Biotechnology and the Future of World Agriculture. London: Zed Books, Ltd. 1991.

16. Cohen JI. Biotechnology research for the developing world. Trends in Biotechnology 1989; 7:295-303.

17. Organization for Economic Cooperation and Development. Biotechnology: Economic and Wider Impacts. Paris: OECD, 1989.

18. Junne G. The Impact of Biotechnology on International Commodity Trade. Ch. 8 In: Da Silva EJ, Ratledge C, Sasson A, eds. Biotechnology: Economic and Social Aspects: Issues for Developing Countries. Cambridge, UK: Cambridge University Press, 1992.

19. Echeverría RG. Assessing the impact of agricultural research. In: Echeverría RG, ed. Methods for Diagnosing Research System Constraints and Assessing the Impact of Agricultural Research, Vol. II. The Hague: International Service for National Agricultural Research, 1990:1-31.

20. Evenson RE. Notes on the measurement of the economic consequences of agricultural research investments. In: Lee DR, Kearl S, Uphoff N, eds. Assessing the Impact of International Agricultural Research for Sustainable Development. Ithaca, NY: Cornell International Institute for Food, Agriculture, and Development, 1992.

21. Ruttan VW. Agricultural Research Policy. Minneapolis: University of Minnesota Press, 1982.

22. Pardey PG, Roseboom J, Anderson JR. Regional perspectives on national agricultural research. Ch. 7. In: Pardey PG, Roseboom J, Anderson JR, eds. Agricultural Research Policy: International Quantitative Perspectives. Cambridge: Cambridge University Press, 1991.

23. Barker R. "Impact of Prospective New Technologies on Crop Productivity: Implications for Domestic and World Agriculture." Paper presented at the Conference on Technology and Agricultural Policy, National Academy of Sciences, Washington, DC, December, 1986.

24. Buttel FH. Sociological Impact. Ch. 23. In: Persley, GJ, ed. Agricultural Biotechnology: Opportunities for International Development. Wallingford, UK: CAB International, 1990.

25. Florkowski WJ, Halbrendt C, Huang CL, Sterling L. Socioeconomic determinants of attitudes toward bioengineered products. Review of Agricultural Economics 1994; 16:125-132.

26. European Commission. Biotechnology and Genetic Engineering: What Europeans Think About It in 1993. Eurobarometer 39.1, Report written for Directorate-General, Science, Research and Development, European Commission, October 1993.

27. World Bank. Agricultural Biotechnology: The Next "Green Revolution"? Washington, DC: The World Bank, 1991.

28. Evenson RE, Putnam J. Intellectual property management. Ch. 25. In: Persley GJ, ed. Agricultural Biotechnology: Opportunities for International Development. Wallingford, UK: CAB International, 1990.

29. Doelle HW, Gumbira-Sa'id E. Joint microbial biotechnological ventures in developing countries: social promises and economic considerations. Ch. 10. In: Da Silva EJ, Ratledge C, Sasson A, eds. Biotechnology: Economic and Social Aspects: Issues for Developing Countries. Cambridge: Cambridge University Press, 1992.

30. Kenney M. Biotechnology and the public sector. In: Sasson A, Costarini V, eds. Biotechnologies in Perspective: Socio-economic Implications for Developing Countries. Paris: UNESCO, 1991.

31. Kaiser H, Tauer L. Impact of the bovine somatotropin on U.S. dairy markets under alternative policy options. North Central Journal of Agricultural Economics 1989; 11:59-73

32. McGuckin JT, Ghosh S. Biotechnology, anticipated productivity increases and U.S. dairy policy. North Central Journal of Agricultural Economics 1989; 11:277-287.

33. Chiou GT, Chen DT, Capps O Jr. A structural investigation of biotechnological impacts on cotton quality and returns. Amer Jour Agr Econ 1993; 75:467-478.

34. National Research Council. Plant Biotechnology Research for Developing Countries. Washington, DC: National Academy Press, 1990.

REPRESENTATIVE TRANSFER OF PLANT BIOTECHNOLOGY WITHIN DEVELOPING COUNTRIES

CHAPTER 5

SOUTH AFRICAN PLANT BIOTECHNOLOGY

Jocelyn Webster and Muffy Koch

Any visitor to South Africa comments sooner or later on the re-markable coexistence of First World and Third World environments in this newly democratized country. This dichotomy provides a rich field for technology transfer to developing communities. Research groups that have striven for years to maintain their position in Western science are now largely turning to the application of science for the up-lifting of disadvantaged people. The driving force for this has been primarily refocusing of government priorities and resources, much of which is channeled into upliftment programs. However, the initial work in plant biotechnology preceded this funding change, and the coexist-ence of business-oriented and community-oriented biotechnology products can be seen in the work discussed here. The products of First World plant biotechnology aimed at crop improvement are easily translated into benefits for developing communities. Only the funding is less readily available for these upliftment programs.

The agriculture sector is the largest sector in South Africa. In 1992 agriculture had an estimated gross-value output of $6.7 billion and employed 10% of the country's economically active population (Zenda Ofir, Foundation for Research Development (FRD), South Africa, un-published data). This sector is served by the Agricultural Research Council (ARC), the CSIR, the Department of Agriculture, universities and private companies with opportunities also existing in horticulture.

PLANT BIOTECHNOLOGY CENTERS

RESEARCH ORGANIZATIONS

The two major research organizations involved in plant biotechnology in South Africa are the CSIR and the ARC. Both obtain government funding and are also required to bring in private research contracts. The twelve divisions at the CSIR carry out research, development and implementation projects in science, engineering and technology. The CSIR spends 8 to 10% of all R&D funding spent in Africa. The divisions focused on food science (FOODTEK) and forest science (FORESTEK) have projects in plant biotechnology. A biotechnology program was established in 1979 at FOODTEK, and the existing technology is considered as one of FOODTEK's four core competencies. The first FOODTEK product to be commercialized will be improved lysine production in bacteria. Projects are required to be centered on a market and linked to a commercial partner. The major thrust in CSIR plant biotechnology programs is the development of molecular markers and screening systems (e.g. DNA fingerprinting) for plant breeding programs, the isolation of novel genes for crop improvement, especially fungal, insect and drought resistance, and the development of transformation systems for specific crops. Projects for commercial partners include the transfer of herbicide resistance into soybean for AgrEvo (Germany) and modification of tomatoes for Mayfords, the genetic manipulation of maize and the improvement of sunflowers for South African clients. FORESTEK's research has centered on early screening systems for breeding programs, improved rooting for seedlings and the development of somatic embryogenesis for propagation of commercial tree varieties, mainly pine and eucalyptus. This group is also investigating local germplasm for valuable secondary metabolites and pharmaceuticals. In collaboration with organic chemists at the CSIR, both groups have investigated biotransformation.

The ARC is the largest agricultural research organization on the continent. Their work is carried out at centers throughout the country, and plant biotechnology is most evident at the Vegetable & Ornamental Plant Institute (VOPI), the Stellenbosch Institute for Fruit Technology (INFRUITEC), Grain Crops Institute (GCI), Plant Protection Research Institute (PPRI) and the Institute for Tropical & Subtropical Crops (ITSC).

At VOPI plant biotechnology has been applied to viral resistance in local ornamental bulbs, cucurbits and potatoes, to proline applications for drought and heat tolerance, synthetic seed production, fungal resistance genes for cotton and tobacco, molecular markers for grasses, grapes, potatoes and alfalfa, *Bacillus thuringiensis* (Bt) genes for cotton bollworm and phosphinothricin resistance in soybean. At INFRUITEC considerable effort has focused on bacterial resistance strategies for fruit trees, viral resistance for granadillas and strawberries and molecular markers for fruit cultivars, and they have recently reported the first transfer of a herbicide resistance gene into a commercial strawberry cultivar. An application has been placed for field trials with the transformed strawberries. Tissue culture procedures for the regeneration and transformation of deciduous fruit trees have been developed in the

INFRUITEC laboratories. The PPRI biotechnology programs are primarily concerned with viral detection. The ITSC and GCI employ plant biotechnology for hybrid production by embryo culture, meristem culture for disease free material, cell transformation and fingerprinting.

UNIVERSITIES

Ten of the 18 tertiary education institutions reporting biotechnology activities have established projects in plant biotechnology.[1] The three largest groups are detailed. The University of Cape Town (UCT), which has world-renowned molecular biology expertise, has concentrated its efforts on antiviral constructs for major viral diseases of maize, including maize streak virus, isolation of Bt genes and modifying protein content in local disease resistant varieties. They also are involved in gene cloning from indigenous medicinal plants and gene synthesis for crop improvement. The University of Natal has strong biotechnology groups using the techniques for studying nitrogen and carbon metabolism, senescence, hormonal regulation, rooting in tree species and secondary metabolite production. The University of the Orange Free State (UOFS) is isolating genes from, and molecular markers for, local Russian aphid resistant wheat varieties.

COMMERCIAL PLAYERS

The two major industries involved in plant biotechnology are the chemical and sugar industries. Both of these industries have set up their own in-house plant biotechnology facilities. The African Explosives and Chemical Industries (AECI) group is linked to ICI, UK and is working on improvements in apples, Bt production and viral resistance in specific crops. The South African Sugar Association (SASA) group is working on genetic mapping, tissue culture, gene characterization and selection and genetic engineering for insect, viral and drought resistance in sugarcane. Other companies in South Africa that are developing or implementing plant biotechnology products are listed in Table 5.1, together with their chosen overseas partners where appropriate.

A recent Arthur D. Little (U.S.) survey[2] suggests that for a company to succeed commercially in biotechnology it need not be the source of new science or enabling technology, but needs to be current in the most advanced science and technology to be able to evaluate the technology critically and to equip itself to apply the technology. In the South African environment it is probably acceptable for biotechnology R&D to be one to three years behind the most advanced work in laboratories in other nations.

CROPS AND GENES OF INTEREST

The major crop in South Africa is maize, which is the staple food of the population. As such, maize production is aimed at human consumption with all the inherent quality and health factors. The commercial

Table 5.1. Selected South African companies involved in biotechnology projects or products with international partners where appropriate

SA Co.	Overseas partner	Product
AgrEvo	AgrEvo	Herbicide resistant maize and soybeans
National Chemical Products (NCP)	Yes	Biocontrol and biofertilizer
Monsanto	Monsanto	Bovine somatotropin
Hadeco	—	Micropropagation of disease free ornamental bulbs
PHI Genetics	Pioneer Hi-Bred	Maize, soybeans, sunflower and sorghum
Sensako	DeKalb Genetics Corp.	Virus resistance in maize Pest resistance in wheat Molecular markers in breeding program
Pannar	Yes	Modification in sunflowers
Carnia	Cargill	Canola and maize
Carnia	Asgrow	Soybeans
Mayfords	Yes	Genetically improved tomatoes
Clark Cotton	Delta & Pine Land Co.	Bt transgenic cotton

maize producers are reliant on seed companies for their cultivars, and most local seed companies have overseas affiliations that provide their plant biotechnology requirements (Table 5.1). The genes available for improving local cultivars are frequently of value to small, rural farmers as well. Transfer of this technology to these disadvantaged producers has proved difficult. Obtaining patented insect and viral resistance genes for developing communities has been unsuccessful. In many instances the most feasible route appears to be to clone local genes using probes based on patented genes. This has been UCT's approach in obtaining a systemin gene following their inability to obtain the gene from a research group. Cloning is an expensive, complex and time-consuming process, but appears to be the best option for using available research funds to transfer crop improvement technology to the rural small farmer.

Wheat, like maize, is farmed using highly developed systems. There is not much evidence of wheat being farmed as a subsistence crop. Sorghum is the next most important cereal crop and is used for both human consumption (it makes a nutritious beer and food) and fodder.

It is a traditional food and has better drought tolerance than maize, which makes it favorable to many small farmers, who could benefit from improved characteristics like insect and fungal resistance, improved drought tolerance and improved brewing characteristics.

Fruit and vegetable crops of great variety are grown in this country. Most of these crops would benefit from improved characteristics such as insect resistance, fungal and bacterial disease resistance, drought tolerance and viral resistance. Even the much publicized improved shelf life in Western tomatoes would be a benefit to both the small growers and industry in this country. Efforts by biotechnologists to obtain these and other genes for testing in local varieties have been largely unsuccessful.

Genes being developed by local centers are mostly aimed at important local problems. These include fungal-resistance genes (FOODTEK, ARC, UOFS), viral-resistance genes (ARC, SASA, UCT), insect-resistance genes (AECI, ARC, FOODTEK, UCT, UOFS), drought-tolerance genes (VOPI) and bacterial-resistance genes (INFRUITEC).

CURRENT RESEARCH VS. NEEDS OF THE COUNTRY

A clear omission in the plant biotechnology research listed above is that although the research projects are targeting important criteria, the results of the research effort will not be available to rural communities. Most of the research projects mentioned in the previous section are linked to a commercial partner, due to tight funding constraints on biotechnology research in this country. To secure return on investment the co-funding companies require a tight control over the research and the use of the genes. This is best obtained with the high-volume, high-value crops and has led to a number of research groups competing for funding on similar or the same projects.

In a small country with limited funding it would seem imperative that a national strategy should guide the effective use of research funds. In plant biotechnology this is clearly the case. According to statistics from the Foundation for Research Development (FRD), which administers most government spending on agricultural research at tertiary education institutions, 45% of 53 academic research groups working in medical science, agriculture and industrial microbiology have plant biotechnology projects (Z. Ofir, personal communication). Universities, the ARC and the CSIR use government funding to develop expertise in plant biotechnology. Competition between all the plant biotechnology groups for high return projects is good in one sense, but also runs the risk of wasting money, by preventing cohesive, well-structured approaches to important problems. We believe that a national strategy designed to coordinate, not bureaucratize, research efforts in plant biotechnology would greatly stimulate this field and efficiently involve all the groups in South Africa that are eager to participate in the technology. Without this policy there will be no coherent setting of priorities, no transfer of commercial genes to noncompeting rural farmers and no policy to guide spending and return on investment. As noted by Zilinskas,[3] governments must create the economic climate conducive to entrepreneurship and risk taking that is needed to establish a biotechnology-based industry.

LAWS AND REGULATIONS

Patent protection is one of the main conditions for ensuring a return on investment, and it plays an important role in a company's decision to invest in agricultural biotechnology. South African patent law, based on the European Patent Convention, prohibits the patenting of plant and animal varieties in principle. Recent modification to European law in Plant Breeders' Rights included protection for "essentially derived varieties." To meet these and other specifications The Plant Breeders' Rights Amendment Bill will be submitted to the South African Parliament in 1994. Plant breeders will have to obtain evidence that a new variety does not differ significantly in the genetic make-up or expression of genes from the initial protected variety. The breeder of a derived variety has the burden of proof to show that his breeding methods do not constitute plagiarism. As such, varieties with single-gene modifications due to genetic engineering may be protected under plant breeders' rights.

South Africa has set up guidelines[4] for the control and release of genetically modified organisms (GMOs). A safety committee, called the South African Committee for Genetic Engineering (SAGENE,) was established in the late 1970s and has played an essential role in the control and release of GMOs. The SAGENE guidelines are amended

Table 5.2. Field trials of transgenic crops approved by SAGENE

Crop	Transgenic trait	Applicant	Year
Cotton	Bromoxynil tolerance	Calgene	1990/1991
Cotton	Bromoxynil tolerance / Insect resistance (Bt)	Calgene	1991/1992 1992/1993 1993/1994 1994/1995
Cotton	Insect resistance (Bt)	Clark Cotton	1993/1994
Cotton	Insect resistance / glyphosate tolerance	Clark cotton	1994/1995
Maize	Phosphinothricin tolerance	Hoechst	1993/1994
Lucerne	Phosphinothricin tolerance	Hoechst	1993/1994
Canola	Phosphinothricin tolerance	Hoechst	1993/1994
Maize	Insect resistance / glyphosate tolerance	Monsanto	1994/1995

as world regulations change, but the committee has no mandate or resources to police the guidelines it has established. Several field trials have been carried out by international companies (e.g. Monsanto) within SAGENE's guidelines (Table 5.2), but these companies, while believing that the guidelines are comprehensive for field testing, are concerned about releasing transgenic material commercially without legislation in place for this.

NATIONAL BIOTECHNOLOGY STRATEGY

Countries that have well-established biotechnology industries are those that have suitable legislation and patent protection and have a structured national biotechnology policy which has provided the funding necessary for R&D, the direction and R&D priorities. South Africa, like many developing countries, has gaps in the required structure to support the development of a biotechnology industry. There is no legislation, patent protection is limited and, most importantly, there is no national policy identifying strategic goals for R&D. Financial support for biotechnology start-up companies is difficult to obtain, especially in the present economic and political climate.

The FRD has attempted to establish R&D priorities for biotechnology in South Africa, but to date there has been little progress by government to define the role biotechnology will play in a future South Africa. In 1991 the Industrial Biotechnology Association of Southern Africa (IBASA) was established by local industry to inform the public about biotechnology to provide a forum for exchange on biotechnology policy and to ensure that industrial biotechnology is afforded a high priority by government and that industrial biotechnology can play a proper role in job and wealth creation in Southern Africa. It remains to be seen the role that IBASA will play in promoting and supporting a biotechnology industry in a democratized South Africa.

To define R&D priorities across the entire spectrum of agriculture biotechnologies is difficult and depends on wide policy goals.[5] If the aim is to rapidly increase protein supply of high nutritional value for both developing and industrialized countries, fish biotechnology could be the priority. If the aim is to replace chemical herbicides, pesticides and fertilizer, increased support for specific plant and microbial biotechnology would be required. If the main aim is to promote the economic adjustment of agriculture, nonfood uses of agricultural crops would be supported. In South Africa where there is no national policy for biotechnology, the R&D priorities have to be market-driven, whilst adjusting our focus to include the needs of the small farmers and developing communities.

DEVELOPING COMMUNITIES

Agriculture in developing communities in South Africa is characterized by farming that is carried out on small stretches of private or communal farmland using a low level of technology. Subsistence farming is less common, as the small farmers sell their crops for cash with which to buy bread, maize meal and other basics (personal communication, A. Pretorius). Agronomy in developing communities frequently competes with valued cattle herds and difficult-to-control goat herds. South Africa's new Reconstruction and Development Programme (RDP) is the basis on which the government will be allocating resources in the next few years. It is aimed at uplifting disadvantaged communities and redressing the inequalities of the past.

Whilst there is some information available on small farmer needs in South Africa, there is no information on the application of biotechnology to address these needs. A study to assess these needs is not a rapid exercise, but it is possible that international aid funding could be obtained to execute such a study, as has been the case in other developing countries.[6] At a national level, such a study would need to involve small farmers, government, nongovernment organizations (NGOs) and local players in plant biotechnology. Our aim here should be to establish small biotechnology-based rural businesses that create wealth and jobs.

CONSTRAINTS ON TECHNOLOGY TRANSFER

A major constraint is the need to balance the application of plant biotechnology for competitiveness, whilst applying the products for upliftment of rural farmers. Other major constraints on biotechnology transfer to South Africa by international organizations are lack of funding, lack of local legislation and fear of risky investment due to unstable politics. Transfer of technology to developing communities is largely impeded by lack of access to developed genetic material and transformation technology. Collaborative agreements between international biotechnology companies and research groups serving developing communities, would greatly alleviate this bottleneck. International agencies can hasten technology transfer by providing technical training, information and equipment.[3] South Africa identifies with Africa in an effort to mobilize plant biotechnology. To this end the FRD has recently organized and co-funded UNESCO's first initiative in South Africa—an all Africa workshop on plant tissue culture.

CONCLUSION

During the next few decades plant biotechnology can make a major contribution to South African economic and social welfare, thereby addressing the aims of the R&D in agriculture, health and nutrition. Biotechnology has the capacity to limit the damaging environmental consequences of some agricultural practices, deforestation and climatic changes. Agrofood biotechnology could make crucial contributions to the health and prosperity of developing communities. It is therefore in the economic interests of the developed and developing countries to devise policies on research, intellectual property, safety and legislation to enable the developing countries to build up their biotechnology capabilities. South Africa has available all of the facilities, many of the structures and much of the expertise needed to develop a strong plant biotechnology industry. A national strategy guiding spending on plant biotechnology would greatly facilitate progress in transfer of this technology into useful products for commercial companies, small farmers and entrepreneurs.

The diffusion of biotechnology into society in the next ten years will be a gradual process without major, destabilizing impacts on social structure or employment. Public perceptions of biotechnology and safety issues remain the major sources of industrial uncertainty and must be addressed by improving communication among governments, scientists, industry, the media and the public.

South Africa has recently won a major political struggle. Can plant biotechnologists and the communities they serve orchestrate a biotechnology breakthrough for this country? We believe they can. The needs' analysis and South Africa's integration into Africa has begun. With continued support and encouragement we expect the momentum to grow.

REFERENCES

1. Biotechnology in South Africa. Pretoria: Foundation for Research Development, 1992.
2. AD Little Inc. The future of molecular biotechnology in agriculture and forestry—a scenario approach. Report to the CSIR, 1993.
3. Zilinskas RA. Bridging the gap between research and applications in the Third World. World J Microbial Biotechnol 1993; 9:145-152.
4. Guideline for the Use of Genetically Modified Organisms (SAGENE). Pretoria: Foundation for Research Development, 1991.
5. Organisation for Economic Co-operation and Development. Biotechnology, agriculture and food. Paris: OEDC, 1992.
6. Bunders JFG, ed. Biotechnology for Small-Scale Farmers in Developing Countries. Amsterdam: VU Univ Press, 1990.

CHAPTER 6

PLANT BIOTECHNOLOGY IN KENYA: OPPORTUNITIES FOR NATIONAL DEVELOPMENT AND TECHNOLOGY TRANSFER

John S. Wafula

INTRODUCTION

The major challenge facing most countries in Africa today is the production of adequate food and industrial raw materials for the continent's rapidly-growing population which is now estimated to be more than 500 million people.[1] Meeting this challenge requires programmatic support for science and technology, particularly in agricultural development.

Agriculture is the backbone of Kenya's economy, providing 30% of the Gross Domestic Product and employing more than 80% of the total workforce. This sector plays a leading role in the provision of food security for the country's 25 million people, a population estimated to be increasing at the rate of 3.4% per annum.

Kenya's agriculture faces many constraints which emanate from long and frequent droughts (e.g. 1992-1994), inadequate planting seed, unsuitable soil types to support plant growth and the presence of diseases and pests as well as a lack of suitable technologies to meet the needs of modern agricultural development. These limitations cause severe shortages in food availability from time to time, which leads to huge importation of basic food-stuffs. The drain on the country's limited resources arising from such importation is massive.

Plant Biotechnology Transfer to Developing Countries,
edited by D.W. Altman and K.N. Watanabe. © 1995 R.G. Landes Company.

The means usually employed to alleviate some of the production hardships are improvement of germplasm for greater product versatility and resistance to environmental factors using conventional agricultural-research methodologies. The emerging developments in biotechnology and its ease of integration with conventional plant-production methods provides greater potential for addressing problems of sustainable crop and tree production.

Although many research institutions in Kenya have been involved in biotechnology research for some time, the landmark of national awareness of agricultural biotechnology was the international "Plant Biotechnology Workshop on Present and Future Biotechnology Research and Application for Kenya" held in Nairobi in May 1989. This meeting set in motion a series of deliberations on biotechnology that led to a number of significant pronouncements by the Kenyan government and various research institutions on the need for promotion of biotechnology to support national development. Subsequent discussions have taken place on the relevance and scope of biotechnology to Kenya taking into account national priorities and capacity. In addition, ongoing institutional activities in biotechnology have been reviewed and potential areas for future national involvement assessed along with developments elsewhere, the latter aimed at determining appropriateness and capacity for acquisition of technologies through the transfer process. As a result of these considerations, Kenya's science and technology policy now incorporates application of biotechnology and related issues of biosafety and intellectual property rights (IPR) as a national strategy in research and development.

PLANT PRODUCTION IN KENYA

MAJOR CROPS AND TREES

Agricultural production is carried out by both large- and small-scale farmers, with the latter group representing the largest proportion of the farming population. Most small-scale farmers cultivate traditionally small holdings of 2-7 ha in both high (>850 mm) and low (<600 mm) rainfall areas in rural Kenya,[2] the largest proportion of the farming being for subsistence food crops.

The country depends heavily on cash crops for export earnings. Farm produce forms 70% of all domestic exports, most of which is accounted for by horticultural crops, coffee, tea, pyrethrum, cotton and tobacco. Coffee is the single most important cash crop in Kenya. It is generally grown as a single crop on an area of 153,000 ha in the higher-rainfall highland areas. The total production in 1990 was 145,000 metric tons. Tea is the second most valuable cash crop. It occupies an area of 84,000 ha for annual production of over 150,000 tons. In recent years, horticultural produce, particularly cut flowers such as carnations and roses, and fruits like mangoes, bananas and apples, have emerged as very significant foreign-exchange earners. These crops, together with coffee and tea, constitute more than 80% of the total agricultural-based export-earning capacity. Sugarcane, cotton and pyrethrum occupy 90,000, 111,000 and 10,000 ha of land and generate about 425,000, 8,000 and 70,000 tons of produce, respectively. These crops contribute very significantly to the local industrial base and to the export economy.

The major food crops include: cereals, particularly maize, wheat, sorghum and millet; roots and tubers, especially cassava, potatoes and sweetpotatoes; pulses such as cowpeas and soybeans; oil crops like sunflower, groundnuts and coconuts; and horticultural fruit and vegetable crops. Maize is the staple food for the majority of Kenyans. It is grown in nearly every part of the country in both high and low rainfall areas. The development of maize varieties for the various climatic and environmental conditions has for a long time been a major preoccupation of maize breeders in the country. In 1993 maize occupied an area of 1.3 million ha for a total yield of about 19.5 million bags.[3] More than 80% of maize in Kenya is grown on a small-scale basis for subsistence. The remainder is produced on a large-scale commercial basis. Of the roots and tubers, cassava, potatoes and sweetpotatoes are the most significant food crops produced. They are basically grown on the small-holder level for subsistence. Potato production is mainly carried out in high-rainfall areas and occupies 87,000 ha with an output of about one million tons. Cassava and sweetpotato are mainly produced in Western, Coastal and Eastern Kenya on about 562,000 and 315,000 ha respectively. The most important domestic horticultural produce includes fruits such as apples, mangoes, pineapples and bananas, and vegetables like brassicas, beans and peas. These are grown in various parts of the country, especially in the high and medium rainfall areas.

Forestry and agroforestry are becoming increasingly important contributors to the agricultural and industrial economy as well as in environmental preservation. The total area under agroforestry in Kenya is not known. The main trees recommended for intercropping with other crops include *Gravillea* spp., *Leucenea* spp. and *Caleadra* spp. Trees such as *Eucalyptus* spp. are planted mainly for timber and environmental protection. The other contributions of forestry and forest products to the economy include paper and paper products, fuel, wattle bark and its products, furniture and wood carvings.

CONSTRAINTS TO PLANT PRODUCTION

Plant production in Kenya is a function of a number of factors including genotype and its position in the farming environment, soil sustainability, moisture availability and disease and pest incidences as well as climatic and environmental conditions under which the plant prevails. In any given area, plant production may face a multitude of these factors which work to reduce both growth and yield.

More than 71% of Kenya is arid and semi-arid lands (ASALs). Temperatures in these areas go well beyond 38°C, and moisture availability on average is less than 600 mm per annum. Rainfall failure is a common feature in the ASALs and is often accompanied by high evaporation, thus rendering plant growth impossible. In addition, soil salinity is a major problem affecting plant production. These constraints have created a need for drought- and salt-tolerant plants for most of Kenya. Thus programs are needed for development of drought tolerance to extend crops, such as maize, beans, fruits, vegetables and tubers, into growing regions beyond the highland areas with regular rainfall.

There are numerous diseases and pests which affect growth and production of various plant species. Table 6.1 shows the important crops and their major pests and diseases. The most common diseases are those of viral, bacterial and fungal origin, while the major pests include stalk borers, tuber moths, aphids, mealy bugs and termites. Most varieties of maize, wheat, sorghum and millet suffer from diseases such as streak virus, leaf-blight, rusts and smut and are affected by the various pests. Major pests and diseases of vegetables and fruits are nematodes, leaf spot, blight, white fly, greening, spider mites and white grubs. Grain legumes suffer common mosaic virus, wilt and rust and are affected by cut worm and spider mites. Industrial crops like coffee, tea, sugarcane, pyrethrum and cotton are destroyed by coffee berry disease, sugarcane mosaic virus, bacterial blight and cutworm, while coffee borer and bollworm are important pests. Roots and tubers are affected by potato blight, feathery mottle virus and cassava mosaic virus as well as mealy bug, weevils and white and green flies.

The development of plant germplasm that is resistant to these menaces using conventional methods has for many decades formed a major preoccupation of the Kenyan plant breeder and pathologist. Plant production in Kenya is under high disease infection pressure that renders provision of sufficient clean planting material at required times of the year very difficult. The demand for such materials is more than double the current supply for most crops. The economic loss arising from lack of suitable plant germplasm, environmental factors, pests and diseases in Kenya has not been accurately quantified but is considered to be very high. Reducing these losses, as well as dependency on existing environmentally unfriendly measures of plant, pest and disease control is an urgent national aspiration.

STATUS OF PLANT BIOTECHNOLOGY IN KENYA

ROLE OF BIOTECHNOLOGY

For Kenya, research in biotechnology is considered to provide opportunities for development and application of new techniques to solve specific problems constraining plant production and sustainable agricultural production. Tissue culture, which apart from being relatively

Table 6.1. Major crops and their important pests and diseases

Crops	Diseases	Pests
Cereals barley, maize, millet, rice, sorghum, wheat	cob rot, leaf blight, maize streak virus, rusts, smuts	aphid, army worm, crickets, cut worm, leaf hoppers, shoot fly, stalk borer, termites
Industrial crops coffee, cotton, pyrethrum, sugarcane, tea	bacterial blight, CBD, sugarcane mosaic virus	boll worm, coffee borer, coffee scale, cut worm, nematode
Vegetables bitter gould, brassicas, carrot, okra, onion, tomato, indigenous	black rot, early blight, leaf spot, wilt	nematode, spider mite, white fly, white grubs
Fruits and Flowers avocado, banana, citrus, mango, pawpaw, passion flower	greening, leaf spot, powdery mildew	nematode, red spider mite
Roots and Tubers cassava, Irish potato, sweetpotato, yam	bacterial wilt, cassava mosaic virus, *Erwinia*, late blight	green fly, mealy bug, weevil, white fly
Grain legumes dry bean, French bean, pigeon pea	bean common mosaic virus, rust, wilt	cut worm, spider mite
Oil seeds coconut, groundnut, simsim, soybean, sunflower	leaf blotch, leaf spot, rust, wilt	boll worm, hopper, spider mite, white fly

easy and inexpensive, provides opportunities for rapid returns to investment. Genetic engineering, including gene mapping and introduction of new genes leading to development of genetically-altered, high-yielding and stress-resistant plants, is a field that holds greater promise.

THE APPROACH

The direction and approach taken by Kenya in developing an agricultural-biotechnology base has been that of defining priorities and assessing capacities through national and institutional committees and planning workshops. At the national level the focal point for biotechnology coordination has been the National Advisory Committee on Biotechnology Advances and their Application (NACBAA),[4] which in 1991 prepared a position outline for a national biotechnology program focusing on a number of key priority areas. The main elements identified by NACBAA include tissue-culture procedures and capabilities for propagation of various trees and food crops and for pathogen elimination, the application of diagnostic techniques for detection of disease, the development of molecular markers for use in plant breeding and selection to transfer useful genes into plants for disease and stress resistance as well as development of methods for biological control. NACBAA also has recommended the establishment of partnerships and collaborative ventures with international groups for promotion of research, technology development and transfer. The identification of specific program areas within the framework of the overall national priorities is the responsibility of national research organizations in Kenya, which take into account their mandates and capacities.

A different, but coordinated approach, to the national priority identification of biotechnology was undertaken in 1993 and 1994 through the support of the Special Programme Biotechnology and Development Cooperation of the Netherlands' Ministry of Foreign Affairs, Directorate-General for International Cooperation. The approach has employed a participatory bottom-up problem-identification and program-formulation process and has identified national priorities for agricultural biotechnology, to include multiplication of disease-free quality seed and planting materials for various food crops and trees as well as development of drought-, pest- and disease-resistant cereals, root and tuber crops and pulses. Food and feed processing, biotechnology scientific training, policy development and appropriate-technology transfer and adaptation are given special attention.

CURRENT STATUS

A number of public research institutions, the universities and a few private firms are currently engaged in agricultural research incorporating biotechnological processes.

The Kenya Agricultural Research Institute (KARI), Kenya's largest agricultural research establishment, is mandated to carry out agricultural research for generation, development and transfer of technological packages in support of the national goals on food production and agricultural export. The Institute has taken a lead in formulating and implementing an agricultural-biotechnology program which incorporates the application of tissue culture in mass-propagation of pyrethrum, flowers, potato and sweetpotato. The Institute has extended its expertise in tissue-culture methods to assist and support collaborative work in flower and seed production and improvement with industries such as Oserian Development Company and the Kenya Seed Company. Although the supply of various materials derived from institutional tissue-culture efforts have not fully met the increasing demands, considerable improvements have been made. Expansion of tissue-culture work is planned as a national strategy to include all major crops and trees that are amenable to the technology.

Regarding the application of genetic manipulation and recombinant DNA technology, KARI is very advanced in this field in relation to animal biotechnology. In contrast, application of genetic engineering in plant research is still lagging behind. The only major advancement made so far in this direction is a strong collaborative venture with Monsanto Company (U.S.), in the development of sweetpotato transformation against feathery mottle virus. Plans are underway to initiate international-collaborative work in the use of molecular markers for development of drought- and insect-resistant maize varieties. In these collaborations some aspects of the development process, such as transformation and regeneration, may be undertaken in the collaborating institutions while field agronomic and multiplication work could be carried out at KARI.

Other public institutions involved in plant-biotechnology research in Kenya include Kenya Forestry Research Institute (KEFRI). Among the programs initiated at KEFRI, the biotechnology research program is rapidly gaining strength in developing reforestation systems, particularly for the ASALs in addition to the promotion of farm forest trees and agroforestry. Tissue-culture and legume-*Rhizobium* technologies have been developed and are being applied for tree species such as *Ocotea usambarensis* (camphor wood), *Gravillea robusta* (silky oak) and *Chlorophora excelsa* (mvule).

Tea and coffee research is undertaken by the Tea Research Foundation at Kericho and the Coffee Research Foundation at Ruiru. Although conventional methods have been successfully applied to produce adequate planting material, new biotechnological methods are being used in the induction of embryogenesis in secondary callus of leaf explants and identification of elite traits.

The national universities, particularly the Horticultural Department of Jomo Kenyatta University and the Department of Crop Science of the University of Nairobi, carry out research on bananas, fruits and vegetables using tissue culture for disease elimination and mass-production.

The participation of industry in research and development in Kenya is minimal. The major argument is that minimal financial returns, resulting from low market demands for commodities, limit investment in research and development, and little financial support or collaborative ventures have been forthcoming in Kenya. Irrespective of the previous trends, there is an increasing call for partnership in research and development between public and private agro-businesses in Kenya if biotechnological efforts are to reap maximum benefits.

The locally-based international agricultural-research centers also carry out research embracing biotechnology. Although such programs do not focus on issues that relate specifically to Kenya, their activities cut across problems that have relevance to the country's agriculture. Thus the research work by ICIPE and ICRAF on food-crop pests address Kenya's needs to a large extent. These efforts require continuous support from the international and national funding agencies as they provide means for strengthening national development and research capacities.

FUTURE AREAS OF PLANT BIOTECHNOLOGY IN KENYA

The national Development Plan for 1989-1993 set a growth rate of 5% in agriculture in order to meet the requirements of the ever-increasing Kenyan population, especially in the ASALs where people are migrating from high- and medium-potential areas. If the population continues to grow at the current rate, it is estimated that there will be 38.5 million people at the turn of the century. Hence food security will remain a major national priority for many years. More emphasis will need to be given to the development of agriculture in the ASALs in an effort to alleviate poverty and to achieve self-sufficiency in food production.

The main focus in the past has been to grow known crops, such as maize, beans, cowpeas, cassava, potatoes and horticultural crops, with emphasis on indigenous drought-resistant varieties such as sorghums, millet, sweetpotato and cassava. Although the potential yield of these crops is high, it is hardly attained under these marginal conditions. For instance the yields of the cereal crops in general rarely exceeds 500 kg/ha, while those of the grain legumes averages 500 kg/ha. In addition, the cultivation of major fruit crops such as mangoes, bananas, pawpaw and macadamia are restricted by lack of adequate moisture and healthy planting materials.

In terms of cash crops, the horticultural industry continues to be one of the rapidly expanding sub-sectors of the agricultural sector. Demand for horticultural products for local and export markets has continued to grow over the years. In 1990 over 4,914 tons of fresh horticultural produce worth $34 million was exported as compared to 3,224 tons in 1970. On the other hand, other major cash crops like coffee, tea, cotton and sugarcane have continued to suffer, mainly due to unavailability of clean planting materials, pests and diseases, and a decrease in production in recent years.

It is envisaged that in Kenya agricultural biotechnology will be applied to enhance salt and drought tolerance to crops so as to withstand the stress conditions in the ASALs. This will entail application of the molecular-markers approach to develop stress-tolerant plant varieties. For food- and cash-crop production in high- and medium-rainfall areas, biotechnology research will be applied to mass production of clean, pathogen-free planting seed and seedlings and plant-genetic alteration to impart resistance to disease agents and insect pests through the methods of genetic engineering. Additional areas of future biotechnology development in Kenya will have to come through technology transfer from more-advanced laboratories.

PLANT BIOTECHNOLOGY TRANSFER

CONVENTIONAL TECHNOLOGY TRANSFER

Farming systems in Kenya vary widely from the decentralized small-scale subsistence farming to the large-scale commercial enterprises. In terms of food security, the small farms, which constitute over 80% of the agricultural sector, are responsible for nearly all the national food requirements. In addition, they contribute significantly towards the export market with respect to small-scale cash-crop production. The export potential and foreign exchange-earning capacity is, however, the domain of large commercial producers.

The strategy for sustainable agricultural growth is to move towards intensification and growth in commodity production, based in part on improved technology. While large commercial firms in Kenya are better placed to exploit modern technologies, links and mechanisms for transfer and adaptation of improved technologies by small-holder agricultural farmers have been weak and require considerable strengthening. Any new technologies meant for the small-scale agricultural sector must be carefully assessed in terms of appropriateness to local circumstances and the rate and capacity for its adoption.

A research/extension/farmer linkage exists in Kenya which, although still in need of strengthening, is employed in the transfer of conventional technologies to the small-holder farmers in the country. It is important that this conventional technological base be maintained and supported as it is likely that biotechnology innovations from public establishments will be developed and released through this system.

Another good avenue of linking products of public research to the farmer would be to develop linkage mechanisms or licenses to the private commercial firms for final-product development and commercialization. A number of companies in Kenya are involved in germplasm development as well as in-country testing of seed, breeding lines and cultivars imported from outside the country. Vigorous initiatives are needed to link the efforts of public institutions with commercial activities addressing productivity constraints related to national problems. In Kenya, this is an area of weakness arising from lack of policies that create strong relationships between public-research institutions and the private sector.

These problems are beginning to be addressed through establishment of contractual mechanisms such as development agreements and collaborative research between public institutions and private-sector firms. Arrangements have, for example, been made between KARI and a private floriculture firm, the Oserian Development Company, to enhance increased flower production through a joint application of mass-propagation using tissue-culture methods.

There is also a growing appreciation of the need for technology development and transfer from public institutions to the consumers through the commercial sector. In such relationships it is important that communication be initiated at an early stage of development.

BIOTECHNOLOGY TRANSFER

The generation of products of biotechnology, including genetically-modified materials, is part of on-going and potential research programs of several institutions in Kenya. We must bear in mind the high costs involved in the application of complex genetic-engineering technologies and we recognize that remarkable advances have been made by developed countries in the generation of technologies and products of modern biotechnology. One important option for Kenya for germplasm transformation would be, in addition to its own developments, to incorporate products of genetic engineering and molecular mapping through collaboration with advanced laboratories elsewhere. To enable and facilitate such exchange, it is important to develop national policies on biotechnology and biosafety and to strengthen structural frameworks for acquisition of biotechnological innovations.

Methods for transfer of technology from outside the country will vary. These may take the form of joint-research ventures in biotechnology between Kenyan research institutions and advanced public and private laboratories in developed countries. A few such linkages are already in place. The Monsanto/KARI sweetpotato transformation work is a noteworthy example of a joint private/public development program which currently enjoys support from the United States Agency for International Development (USAID). Another example is a joint KARI/Seiberdorf Laboratories (Austria) program on cassava mutagenesis and breeding under the support of FAO/IAEA. In addition, the USAID-supported Agricultural Biotechnology for Sustainable Productivity (ABSP) program and the International Service for the Acquisition of Agri-Biotech Applications (ISAAA) have launched collaborative programs with KARI under which joint studies in genetic-engineering methods and training in biosafety and IPR-related issues have been initiated.

Other opportunities for technology transfer can be envisaged between public institutions and international biotechnology programs. For example, KARI's potato program works in close collaboration with CIP and has developed a mechanism to facilitate international testing and release of potato varieties developed in CIP laboratories. Similar linkages could be developed with other crops such as cereals and pulses, which have not attracted large-scale private-sector investment but which could be supported through the international donor community.

Development of a cooperative biotechnology program between Kenya and the Netherlands within the framework of the Special Programme Biotechnology and Development Cooperation is under discussion. The program focuses on small-scale agriculture and aims to contribute directly to poverty alleviation through support for research, training, priority-setting and policy development. These and similar bilateral developments are additional mechanisms that could supplement the country's financial support for biotechnology research and development, facilitate access to modern technologies and provide training opportunities to support national capacity building.

CAPACITY BUILDING AND POLICY ISSUES

Although major initiatives are underway to enhance technology acquisition through importation and transfer, various complications still exist that hinder the effectiveness of the process. First is the general lack of awareness and knowledge of what goes on. Cohen[5] has attempted to address this gap by preparing a comprehensive index of international biotechnology programs providing opportunities for collaboration and program funding. Organization of this information is a major achievement as it provides national programs with insights on global developments in biotechnology.

Another major problem is lack of adequate expertise and infrastructure to provide a framework for reception, evaluation, multiplication and transfer of technology. Technology-transfer programs in developing countries such as Kenya should include strong elements of capacity building through support of scientific and technical training and infrastructure building.

Of greatest implication, however, are national policies on biosafety and IPR as related to technology development, acquisition and application. To appreciate the opportunities offered by biotechnology, there should be put in place structures and policy frameworks for advancement of biotechnological processes and innovations. This should include the establishment of principles and concepts for the safe handling of materials derived from genetic-engineering work. Like many developing countries, Kenya has just begun developing regulatory mechanisms and biosafety guidelines on the safe use of biotechnology products.

This is a major step forward as it aims to enhance Kenya's capacity to fully realize the impact and contribution of biotechnology to sustainable agriculture.

REFERENCES

1. Winrock International. Assessment of Animal Agriculture in Sub-Saharan Africa. Marrilton, Arkansas: Winrock International, 1992.
2. Durr G, Lorenzl G. Potato Production and Utilization in Kenya. Lima: International Potato Center, 1980.
3. Emongor RA. Crop Production in Kenya with Reference to Tissue Culture Technology: Feasibility Study. Nairobi: Kenya Agricultural Research Institute, 1994.
4. NACBAA. National Advisory Committee on Biotechnology Advances and Their Applications. Nairobi; Ministry of Research, Science and Technology, 1992.
5. Cohen J. International Biotechnology Service. International Initiatives in Agricultural Biotechnology. A Directory of Expertise. The Hague: ISNAR, 1994.

AGRICULTURAL BIOTECHNOLOGY IN INDONESIA: NEW APPROACH, INNOVATION AND CHALLENGES

Sugiono Moeljopawiro and Ibrahim Manwan

INTRODUCTION

Over the past two decades Indonesian agriculture has developed consistently, and the productivity of agriculture has increased significantly. This increase in agricultural production has been obtained through conventional methods. However, this rate of production increase can no longer meet the need for food and raw materials for industrial purposes. New methods that are more effective and efficient than conventional breeding are therefore needed.

Recent developments in cell and tissue culture,[1-3] molecular biology and DNA technology[4] have opened up a new horizon in further advancing agricultural research in Indonesia. These techniques hold out the possibility of relieving the present biotechnological constraints on agricultural production. Biotechnology, which is relatively new, is expected to provide a better way of overcoming these production constraints. The integration of modern biotechnology into conventional agricultural research will have better comparative advantages to sustain agricultural production in the country.

The Agency for Agricultural Research and Development, Ministry of Agriculture has a strong base for agricultural biotechnology applications to enhance agricultural research for increasing production, productivity

Plant Biotechnology Transfer to Developing Countries,
edited by D.W. Altman and K.N. Watanabe. © 1995 R.G. Landes Company.

and utilization of agricultural wastes. In order to achieve more efficient and effective use of available resources, the agricultural biotechnology research program within the next 5-year development plan will focus on areas in which biotechnology will most likely provide new breakthroughs in agricultural development.

PERSPECTIVE ON AGRICULTURAL DEVELOPMENT

OBJECTIVE AND STRATEGY

The objectives of agricultural development are:

1. To improve quality and stability of food self-sufficiency;
2. To increase production and quality of agricultural products for both national consumption and export;
3. To diversify agricultural commodities for market expansion and improvement of job opportunities;
4. To improve farmers' income through increased farm productivity and added value of agricultural commodities; and
5. To improve farmers' capability and participation in the village unit cooperative and farm community.

To achieve these objectives, there are three basic agricultural development strategies: (1) orientation of farmers' welfare toward increased production; (2) improvement of cooperation among farmers and the public and private sectors; and (3) improvement of agriculture's role in achieving integrated and harmonized regional development.

SOME MAJOR ACHIEVEMENTS

In 1988, of the country's 2 million square km of land, 112,000 square km was devoted to forest lands—which include reserve forest, production forest, and parks—and a half million square km for various agricultural activities (Table 7.1). Out of agricultural land, 38% was devoted to woody plants, 17.1% to estate crops, 23.5% to seasonal crops and bareland under rainfed dryland conditions, 15% to wetland rice, 5.6% to steppe pasture and only 0.4 and 0.3% left to dikes and freshwater ponds, respectively.[5] These figures indicate the great potential for the production of food and estate crops.

Table 7.1. Land utilization for agriculture and forestry, 1988

Type of utilization	Thousand ha
House compound	4,836
Rainfed bareland & garden	12,235
Steppe pasture & grass land	2,922
Temporarily unutilized land	9,508
Small-holder tree plants	19,688
Estate tree plants	8,910
Paddy field	7,774
Coastal dike	211
Fresh water pond	121
Protection forest	30,316
Parks and reserved forest	18,725
Limited production forest	30,525
Permanent production forest	33,867
Conversion forest	30,537

Source: CBS[5]

Many of the seasonal barelands, which have been used for animal grazing, are potentially available for crop production. Agriculture plays a significant role in the Indonesian economy. Currently, 55% of the population depend on agriculture for principle earnings. This sector accounts for 20% of the GDP and over 60% of the value of nonpetroleum exports. Over the past 2 decades agricultural output has grown about 4% per annum. The rapid increase of rice production has provided a large part of this growth and accounted for more than 40% of agricultural output, land use and employment. Rice production increased from around 12 million tons in 1969 to over 30 million tons in 1991. Within the agricultural sector, food crops including vegetables and fruit are the most important sub-sector, followed by industrial and estate crops, livestock and fish production. The adoption of improved technology and management has been one of the major factors contributing to the increase in the production of major agricultural commodities, by several fold, during the past 11 years (Tables 7.2 and 7.3).

The impressive growth performance of the agricultural sector contributed substantially to an achievement of Indonesia's development objective, not only for food security and low and stable prices, but also for generation of employment and foreign exchange earnings. The sector's most notable contribution has been in significantly reducing poverty from 60% (1976) to 15% (1991) of the total population.

Through the application of improved production technologies and management of inputs, Indonesia has made major advances in food crops production. For example, Indonesia was the largest rice importer for many years but became self-sufficient for rice in 1984. Rice and soybean production in 1987 were 40,078,195 and 1,160,963 tons with the average yields of 4.5 and 1.2 t/ha, respectively. Despite the progress made in rice and soybean production, continuous research is a must in order to sustain self-sufficiency in food.

Corn is an important source of carbohydrates and therefore a good alternative to rice. A portion of the corn produced is used for animal feed. Corn also fits well in various cropping systems,[6] thus contributing to the farmer's income. The projected demand for corn in future years will continue to increase. Corn demand for food and for feed or industrial purposes (in millions of tons) is projected to be 3.315 and 2.950 in 1995, and 3.576 and 4.936 in the year 2000, respectively.

Sweet potato and cassava are important palawija (nonrice food crops), after corn and grain legumes. Among the tuber crops grown in Indonesia sweet potato is the second most important after cassava. In the last decade the area planted to sweet potato was 250,000 hectares per year with a total production of more than 2 million tons. The area planted to cassava was 1.3 million hectares with a total production of over 14 million tons per year. These root crops usually are consumed either as a main staple food or as a rice substitute. They are consumed as vegetables, and are used in snacks, desserts and salads. Their young leaves can also be used as a vegetable.[7]

Table 7.2. Production of food crops, vegetables and fruits, 1980-1989 (1,000 tons)

Commodity	1980	1981	1982	1983	1984	1985	1986	1987	1988	1989
Rice	29,652.0	32,774.0	33,584.0	35,303.0	38,136.0	39,033.0	39,727.0	40,078.0	41,676.0	44,725.0
Wetland rice	27,993.0	30,989.0	31,775.0	33,294.0	36,071.0	37,027.0	37.740.0	37,970.0	39,316.0	42,371.0
Dryland rice	1,659.0	1,785.0	1,808.0	2,009.0	2,119.0	2,005.0	1,987.0	2,109.0	2,360.0	2,354.0
Corn	3,994.0	4,501.0	3,325.0	5,087.0	5,280.0	4,329.0	5,920.0	5,156.0	6,652.0	6,193.0
Cassava	13,774.0	13,301.0	12,988.0	12,103.0	14,167.0	14,057.0	13,312.0	14,356.0	15,471.0	17,117.0
Sweet potato	2,007.0	2,093.0	1,675.0	2,213.0	2,156.0	2,161.0	2,090.0	2,013.0	2,159.0	2,224.0
Peanut	470.0	474.0	——	——	5,325.0	5,280.0	642.0	533.0	589.0	619.0
Soybean	653.0	704.0	521.0	536.0	769.0	870.0	1,227.0	1,161.0	1,270.0	1,315.0
Spring onion	76.3	79.4	71.0	71.6	107.7	144.9	150.7	na	166.3	243.9
Shallot	217.7	176.0	156.4	283.8	295.1	361.0	382.1	na	379.4	399.5
Potato	230.4	216.7	157.7	250.0	371.5	372.8	446.3	na	418.2	559.4
Radish	31.1	21.0	17.9	21.2	21.7	22.3	26.3	na	22.1	28.0
Cabbage	323.0	349.0	316.1	391.3	584.1	665.4	820.4	na	771.3	926.1
Mustard green	102.0	123.5	122.9	134.8	153.0	189.4	212.4	na	230.5	282.2
Carrot	42.8	54.8	50.5	53.0	54.2	71.3	108.4	na	132.4	192.6
Beans	58.0	43.4	25.3	91.8	71.7	54.9	77.1	na	52.5	69.8
Avocado	46.4	71.6	51.5	45.7	58.1	62.8	72.2	na	62.5	61.3
Orange	311.0	465.7	342.7	493.0	538.5	485.2	574.3	na	445.0	268.6
Lansium	79.5	101.0	54.5	58.4	69.6	53.0	75.7	na	101.5	47.0
Durian	153.1	130.3	178.6	124.5	146.9	150.6	200.2	na	193.2	139.2
Mango	325.2	308.6	424.2	447.9	442.2	416.4	415.0	na	532.0	445.0
Papaya	315.2	322.9	294.7	240.5	296.0	255.4	315.8	na	346.0	323.0
Salacca	57.1	56.8	45.5	52.0	46.4	94.9	87.6	na	115.0	97.3
Pineapple	180.5	214.4	296.6	322.9	471.6	308.8	809.2	na	358.0	215.4
Rambutan	113.4	144.7	139.8	91.0	108.3	93.3	199.2	na	227.0	146.9
Banana	1,976.8	1,058.3	2,032.9	1,781.5	1,991.7	1,908.6	2,079.1	na	2,809.0	2,192.1
Sapodilla	47.3	50.4	55.6	40.9	50.6	51.3	54.8	na	52.4	53.0
Water apple	484.8	380.8	126.3	——	192.0	199.0	264.1	na	306.5	342.8

na = not available   Source: CBS[5,12-14]

Table 7.3. Production of estate and small holder industrial crops, 1981-1990 (1,000 tons)

Commodity	1981	1982	1983	1984	1985	1986	1987	1988	1989	1990
Estate Crops										
Rubber	300.8	301.8	308.5	313.7	320.8	332.1	327.3	425.0	317.2	310.1
Palm oil	752.3	833.8	891.4	1,079.5	1,159.1	1,195.6	1,341.0	1,449.0	1,773.0	1,809.0
Palm kernel	133.4	149.2	156.6	229.9	238.3	249.2	289.1	318.1	386.3	402.4
Tea	85.0	73.6	88.7	102.1	105.1	98.4	100.7	104.2	122.1	125.4
Coffee	23.4	——	16.8	25.7	21.2	26.7	20.8	28.9	31.4	1.4
Cinchona	1.2	1.5	1.9	1.3	2.2	2.8	3.1	2.6	1.5	2.9
Sugar-cane	1,197.3	1,608.7	1,572.1	1,500.0	1,766.5	2,012.9	2,176.0	2,374.0	2,071.0	2,087.0
Tobacco	9.9	11.3	9.1	9.1	9.1	6.5	2.3	4.3	4.1	3.6
M. hemp	0.3	0.3	0.4	0.4	0.3	0.3	0.5	0.4	0.5	0.5
Cocoa	11.4	13.2	12.6	21.8	24.8	21.2	21.6	33.5	38.0	38.0
Rosella	10.1	4.6	9.7	5.9	6.1	19.1	——	——	——	0.8
Small Holders										
Rubber	642.3	585.6	673.6	715.4	732.8	763.2	795.2	838.9	859.1	927.0
Coconut	1,764.6	1,587.2	1,590.2	1,737.5	1,875.2	2,090.9	2,054.0	2,117.0	2,238.0	2,263.0
Coffee	290.4	262.2	287.2	303.4	291.5	334.2	367.8	362.3	397.0	416.1
Clove	28.8	23.4	40.4	42.7	42.7	53.3	76.7	77.9	58.3	90.8
Kapok	46.3	48.5	49.0	48.9	50.6	52.6	52.3	55.7	56.3	37.3
Sugar-cane	913.7	1,373.0	1,248.5	1,393.4	1,379.2	1,416.7	——	——	——	——
Tobacco	99.8	96.9	100.3	82.6	153.4	159.0	109.7	112.6	111.1	123.8
Tea	23.8	16.5	22.9	24.0	27.2	31.1	25.4	25.6	26.3	26.8
Cocoa	1.4	3.9	5.4	6.2	9.0	8.8	25.8	39.8	65.5	69.8
Cashew-nut	11.4	16.8	18.0	19.4	21.1	30.2	24.0	23.2	27.4	34.4
Nutmeg	18.4	14.9	14.5	17.9	14.2	16.2	15.3	14.6	14.8	15.8
Cashie-vera	13.5	12.9	16.9	20.4	20.8	20.0	26.4	25.4	25.8	27.3
Pepper	39.8	39.6	45.8	43.0	40.0	39.6	49.3	56.2	60.4	63.2
Palm oil	1.0	3.0	3.5	4.0	43.0	89.8	165.2	266.9	438.7	525.9
Palm kernel	0.1	0.4	0.5	0.6	5.2	11.3	29.9	53.2	87.7	105.3
Cotton	13.7	12.6	13.2	11.2	13.4	52.3	17.5	7.2	13.3	33.3
Citronella	0.5	0.5	0.6	0.4	0.4	0.2	0.2	0.2	0.1	0.2
Castor seeds	0.7	0.7	0.4	0.5	3.6	1.6	1.4	1.1	1.1	1.8
Vanilla	0.7	0.5	0.6	0.5	1.0	1.8	1.8	2.4	3.2	3.4
Rosella	7.5	9.7	7.1	5.8	6.7	6.5	22.2	13.4	13.3	13.3

Source: CBS[5,12-14]

The use of varieties resistant to major insects and diseases for integrated pest management (IPM) has tremendously reduced the amount of pesticides used without affecting the level of yield per hectare of rice. This is a significant contribution towards minimizing detrimental effects on the environment from the intensive use of pesticides.

AGRICULTURAL PRODUCTION CONSTRAINTS

A great number of problems confront the efforts to increase the production of various commodities. These constraints can be physical, biological or socio-economic in nature. Although agricultural production in some parts of Indonesia has benefited from the use of improved technologies, low yields still persist in certain growing areas due to wide environmental diversity and other reasons. This situation negates the value of general recommendations for varieties and management practices, and hence, requires a location-specific approach. Breeding new varieties is extremely complex because there is a need for a range of varieties which are also resistant to pests and diseases and tolerant to environmental stresses, to fit widely varying agricultural production systems.

The intensity of agricultural production, particularly in rice, brings about the build-up of pests and diseases, which becomes the bottleneck to higher productivity and causes yield instability. These include diseases caused by viruses, bacteria and fungi, insect pests such as brown planthopper and rice stemborer and rodents. Through conventional breeding, breeders face difficulty in obtaining sources of resistance to insect pests, diseases and environmental stresses. It is anticipated that techniques in molecular biology will be used to isolate genes, proven to be effective against various pests and diseases, and to construct new genes such as coat-protein genes. Maintaining food self-sufficiency and improving the status of human nutrition are great challenges that need special efforts to be successful. Agricultural development is facing a high population growth rate (1.9%), reduction of fertile agricultural land, outbreaks of insect pests and diseases, declining productivity and natural disasters such as floods and drought.

IMPROVEMENT OF YIELD POTENTIAL

Increasing yield potential, through affecting the fundamental physiological processes governing plant growth, may be difficult to achieve in the near-term through molecular or cellular techniques. However, it is expected that substantial increases in yield through improvement of individual traits via biotechnology may be possible. The analysis of yield gains due to breeding over the past several decades shows that genetic contribution to yields is primarily the result of added disease, insect and environmental-stress tolerances, all of which can be further enhanced through the application of biotechnology.

Genetically, improved crop varieties offer the most cost-effective means of increasing yields. Past experience has shown that crop yield has been raised significantly through genetic improvements without incurring the risks that go with greater use of chemical fertilizers and pesticides. However, sources of resistance to important pests (rice stemborer and cornborer) and diseases (rice tungro virus and potato early/late blight) are not easy to obtain and introduce into the existing cultivars conventionally. Therefore, advanced technology must be used to enable the maximum impact of the crop improvement program for increased crop production.

High yield always ranks at the top of any farmer's and/or breeder's list of important traits. The genetics of grain yield is very complicated, but it is certainly a heritable trait. Furthermore, indirect evidence indicates that most economically-useful genetic yield gains are due to successive elimination of yield constraints.

Soybean is the second most important food crop in Indonesia for food, feed and industrial use. Therefore, it receives high priority in the overall national food production scheme. Soybean imports increased from 20,000 tons in 1975 to more than 600,000 tons in 1989. There are several constraints in soybean production at the present time, including low and unstable yield, lack of suitable improved varieties and high incidence of insect pests and diseases as most of the existing varieties do not have adequate resistance.

RESILIENT AGRICULTURE

Agricultural development focuses on achieving a resilient agricultural sector which would be characterized by the following four major criteria: (1) it should utilize natural resources optimally for the prosperity of the people; (2) it should be able to solve problems and surmount challenges which may arise in many forms such as prolonged drought, floods, insect and disease outbreaks and an unexpected decline in market prices; (3) it needs to adapt its production systems to changes in demand and technology; and (4) it should play a positive role within national and regional development programs in producing agricultural commodities, increasing people's earnings and expanding employment opportunities.

BIOTECHNOLOGY FOR SUSTAINABLE AGRICULTURAL PRODUCTION SYSTEMS

POLICY OF THE GOVERNMENT

In the early 1980s biotechnology started to attract the attention of Indonesian scientists. They were convinced that biotechnology could give a substantial contribution to the national socio-economic development. It will not only give better, cheaper and quicker solutions to many production constraints but in some instances it may be the only way to offer a possible solution. The self-sufficiency in rice achieved in 1984, can only be maintained in the long run, by the availability of improved rice varieties that are better adapted to different ecological conditions with resistance to insects, diseases and environmental stresses.

By the end of 1988 an assessment was made of all institutions engaged in biotechnology to determine their stage of development. This assessment showed that there was a wide range in their status. Following this three centers were selected to be the center for a biotechnology research network.

1. The University of Indonesia, Jakarta for medical biotechnology;

2. Central Research Institute for Food Crops (CRIFC), Ministry of Agriculture for agricultural biotechnology (In 1993, the Central Research and Development for Biotechnology (CRDB) of the Indonesian Institute of Sciences was assigned as the second center for agricultural biotechnology); and

3. The Agency of Assessment and Application of Technology for industrial biotechnology.

The government of Indonesia has placed a high priority on the application of biotechnology for the development of agriculture. A great deal of expectations have been placed on the role of biotechnology in achieving and maintaining sustainable agricultural production through increasing productivity, stability, sustainability and equitability of agricultural production schemes.

Biotechnology Laboratories of CRIFC and of CRDB, which serve as the centers for the agricultural biotechnology network, have the following objectives:

- to coordinate and to develop agricultural biotechnology of crops, fisheries and livestock;
- to develop a national agricultural biotechnology network which includes research institutes, universities and the private sector; and
- to develop an international collaborative network with advanced biotechnology laboratories of developed countries and IARCs.

POTENTIAL USE OF BIOTECHNOLOGY

The exploitation of the potential use of biotechnology is vitally important, for both intensification of agricultural production and extension of areas under cultivation, including less favorable environments in rainfed, upland and tidal swamp areas. The food production system in such fragile environments will lead to ecological problems that may seriously affect the basis of productivity. Considerable attention should therefore be paid to other technological breakthroughs. Biotechnology may be such a breakthrough; indeed, it could provide an answer to these changing needs and offer possibilities to arrive gradually at the required level of additional production in a manner acceptable to society.

FUTURE GROWTH OF AGRICULTURE

The future growth of food production will depend on increased cultivation of new land and increased productivity. The increase in cultivated areas will still play an important role in the next decade. However, expansion of new cultivated land for food crops will be limited due to population pressure and continued conversion of fertile agricultural land to other uses. Thus, the growth of food production in coming decades will have to result from increased productivity through the use of a new set of highly-productive technologies, such as new crop varieties, fertilizer, irrigation water, control of pests and diseases and reduction of post-harvest losses.

SUSTAINABLE PRODUCTION AND ECOLOGICAL CONSIDERATIONS

Aside from increased food production, there is still a growing public concern regarding sustainability of agricultural production, environmental degradation, declining productivity in specific sectors, outbreaks of insects and diseases, alleviation of poverty and under-employment. Productivity gains from existing, improved technologies are likely to come in smaller increments than in the past, while the demand for more supply of food, feed and raw materials for industry is increasing. High and efficient use of inputs derived from biotechnology application to sustainable agriculture development becomes a major challenge for the future.

As we look further in the coming years, there is a growing concern about the impact of resource and environmental constraints that may seriously impinge on the capacity to sustain food-crop productivity. More attention and careful analysis are needed to minimize detrimental effects of the intensive production schemes on the sustainability of the food production system.

To meet these challenges and to achieve and maintain sustainable agricultural production systems, four important areas should receive further attention where biotechnology may play a significant role. These are methods that could: continue to improve yield potential or productivity; breed resistance against major insect pests and diseases; minimize detrimental impacts of pesticides and fertilizers on the environment; and increase efficient use of water.

Because biotechnology is still relatively new in Indonesia and means many different things to different people, it is helpful to assess its potential use to strengthen and enhance agricultural development. The following are the key questions to identify the problems and needs in relation to the use of agricultural biotechnology:[8]

1. What are the major problems to be solved in the country to increase agricultural productivity?
2. What new products or processes are needed to solve these problems?

3. Do these products and processes exist, or are they being developed elsewhere?
4. If the required products or processes exist elsewhere, are they available for transfer?
5. If the required products and processes do not exist, do they need to be developed specifically in the country of need?
6. If the required products and processes need to be developed, what is the most efficient way to have them developed, and where?

Although biotechnology has been placed as the top priority, it does not mean that the development of a biotechnology program in Indonesia is without its challenges. The first of these is the need to focus on the agricultural problems which may be addressed through biotechnology and the products and processes needed to solve them, rather simply on the technology itself. There needs to be a clear strategy to identify problems that biotechnology is likely to solve. A second challenge is to pay very close attention to the market for agricultural biotechnology research products. This market consists of agricultural producers, buyers and consumers. Thirdly, is to forge the strongest possible linkages between public-sector biotechnology research activity and the private sector, as well as universities.

MAJOR FACTORS AFFECTING BIOTECHNOLOGY DEVELOPMENT

Biotechnology development in Indonesia will be determined by four major synergistic factors. These are government policy, external support, private-sector and farmers' involvement and improved techniques (Fig. 7.1).

Government Policy

The Indonesian government has declared that biotechnology has been given top priority in the national development. Several steps have been undertaken to promote the development of biotechnology in Indonesia.

External Support

Major external supports considered to play a significant role in biotechnological development include: resource development, a collaborative network and regulatory procedures, such as for biosafety regulation, and intellectual property rights.

Private Sector and Farmers' Involvement

When biotechnology research activities start generating improved products, this is expected to be followed by a scale-up of production. In this regard the participation of the private sector is vitally required. Adoption of biotechnology products by farmers as one of the end-users is equally important.

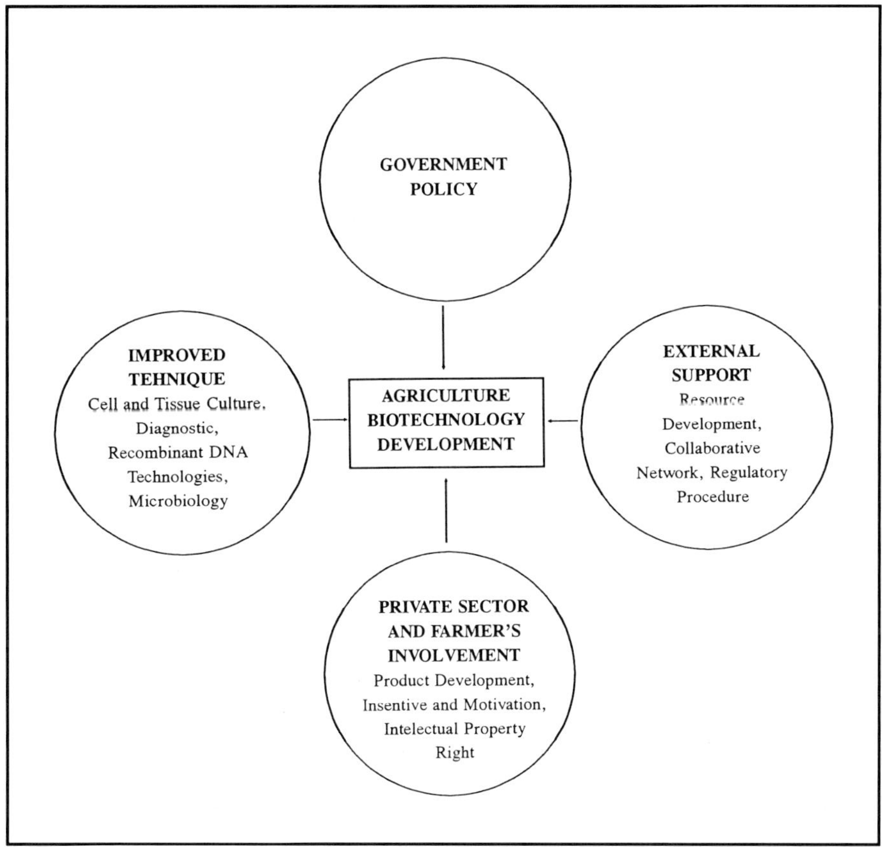

Fig. 7.1. Factors affecting agricultural biotechnology development.

Improved Techniques

Continuous development of new techniques in cell and tissue culture, recombinant DNA applications, diagnostics and microbiology is required to provide backup support for sustainable development of biotechnology in the country.

COLLABORATIVE RESEARCH AND PARTNERSHIPS

COLLABORATIVE MECHANISMS

Research collaboration has become an effective mechanism to speed up the generation of new improved technology and information and to enhance national research capacity. The Central Research Institute for Food Crops collaborates with a wide range of national research institutions, international research centers and donor agencies. A great amount of achievement and progress, in terms of the generation of improved technologies, provision of technical assistance and research facilities, training of research staff, joint research activities and exchange of scientists, has been accomplished through collaborative mechanisms.

Over the last several years CRIFC has established joint, collaborative activities with national and multinational private companies operating in Indonesia through contract research or other mechanisms of collaboration. These collaborations have generated many synergistic benefits, which enable problem-solving more easily and in relatively less time and make possible the movement of important innovations to the end-users. The form of collaboration depends on the subject, the capacity and the objective of the research partners and on the available resources. The major areas covered in these collaborations include varietal development, agricultural machinery, use of agricultural chemicals and agricultural biotechnology.

In addition, collaboration with the International Agriculture Research Centers (IARC) has a long history and has generated a great deal of technologies, information and trained staff which are important in enhancing agricultural development in the country. CRIFC has been receiving support from the Rockefeller Foundation to enhance rice biotechnology. Similarly, CRIFC has just joined the Asian Rice Biotechnology Network funded by the Asian Development Bank (ADB).

The U.S. Agency for International Development (USAID) has consistently encouraged Indonesia to seek out new and innovative ways to work jointly with the private sector on biotechnology research, including cost sharing and creation of public- and private-sector collaborations. USAID is providing assistance to the Central Research Institute for Estate Crops in carrying out a tissue culture project with Indonesia Plantek International. These joint activities have a real potential to benefit the government, the private sector and consequently the farmers. The establishment of the Agricultural Biotechnology for Sustainable Productivity (ABSP) project is a good example of this new innovative approach in which public and private sectors from Indonesia and U.S. have come to an agreement to work on agricultural biotechnology. It is expected that through this project the U.S. and Indonesia can create a means to foster productive and mutually beneficial biotechnology cooperation.

Japan International Cooperation Agency (JICA) has a long history of manpower and infrastructure development of Indonesian agricultural research and development. Tissue culture research facilities have been placed in Lembang Horticultural Research Institute for the production of disease-free planting material of potato, in support of biotechnology research.

Current collaborative research between the Australian Centre for International Agriculture Research (ACIAR) and Indonesia in biotechnology have been on *Rhizobium* and *Pseudomonas solanacearum* and on the development of peanut cultivars resistant to peanut stripe virus (PStV) through the use of a viral coat-protein gene.

STAGES OF TECHNOLOGY TRANSFORMATION

Indonesia, as a member of regional and international networks, is fully aware that there is an interrelationship between national development and development of science and technology. The role of science and technology will become more apparent with the increasing capacity of the nation to pass the transformation period from an agrarian-based economy to industrialization.

The transformation itself, as Prof. Habibie, the State Minister for Research and Technology stated on many occasions, takes place in four overlapping stages. The first and most basic is the stage of technology transfer through license production or use of existing production and management technologies in the production of goods currently in the market. The second is the stage of integrating already-existant technologies in the design and production of a completely new product that is not yet in the market. The new element is an aspect of creation. The third is the stage of technology development per se. The existing technology is developed further. The motivation to undertake this stage is the opportunity to produce a new product for the future. The fourth is the stage of large-scale basic research to support the first three stages and to defend the technological superiority already attained.

Biotechnology, which has been designated as a qualifying technology, will not only produce revolutionary changes in agriculture, but also will have a wide socio-economic impact. The successful passage through these four stages will very much depend on the ability to improve national research capacity, which can be enhanced through partnerships, collaborative efforts and efficient research networks.

BIOTECHNOLOGY FOR AGRICULTURAL DEVELOPMENT

COMPETITIVE RESEARCH GRANT

Realizing that research and development in biotechnology require a tremendous amount of resources, in 1991 the National Development Planning Board (Bappenas) began its survey on human resources and facilities available for biotechnology research in Indonesia. The data obtained were then used as the basis for launching the so-called "Riset Unggulan Terpadu" (Excellent Integrated Research) grant program.

This program will provide funding for research in biotechnology and other sciences to the participating research organizations. Scientists can submit proposals for research focusing on priority issues, with grants being awarded by an evaluation committee. This committee, in coordination with the Ministry of Research and Technology and the National Research Council, has been organized in such a manner as to ensure an unbiased assessment of each proposal while maintaining a strict allegiance to the needs of the national development program in Indonesia.

Following the initial screening, the coordinator will distribute proposals for peer review by qualified research scientists who have experience in Indonesia or in countries with characteristics similar to Indonesia. Proposals that have moved successfully through this screening will be further scrutinized for their scientific merit and appropriate scope, time frame, pertinence to the needs of the community served by the research grants program and the budget's acceptability given limits of funding.

This program has now been going on for two years. The progress of each research project has been monitored every year and, consequently, those with no indication of success are no longer funded. A new grant system called "Riset Unggulan Kemitraan" (Excellence Partnership Research) will be started in 1995 in order to accelerate the commercialization of products from this program. In this program, proposals submitted should be jointly developed by public institutions and the private sector in order to scale-up the production from laboratory level to commercial level.

Another innovative approach used by the ABSP involves the participation of the private sector in both countries for the project. This will ensure rapid transfer of technology from well-funded private companies in the U.S. and private companies in Indonesia. Privatization of research is the current agenda for research systems in Indonesia, and the involvement of ABSP in this endeavor will hopefully accelerate the process.

CONSEQUENCES OF BIOTECHNOLOGY APPLICATION

The capacity of Indonesia to appropriate and adapt biotechnology to its special needs will determine the extent to which the technology will contribute to agricultural production. The contribution of biotechnology will also be determined by the capacity of the country to minimize possible negative effects of biotechnology applications. The following strategies may be used to counter these negative effects and to strengthen indigenous biotechnological capacity.

Access to genetic resources, the very building blocks of biotechnology, is of crucial importance in any attempt to appropriate biotechnology. Most of the diversity of genetic resources originates in developing countries. It is estimated that over 90% of all germplasm monitored by the International Board of Plant Genetic Resources (IBPGR) came from developing countries. Only 15% of that germplasm however, was stored directly in the genebanks of developing countries.[9] Efforts are now underway in the Food and Agriculture Organization (FAO) to guarantee free access to these resources. Although these initiatives are not aimed directly at biotechnology research, they are in fact of utmost importance for any biotechnology program in developing countries.

As we recall in the past, IARCs led the Green Revolution which impacted the rural poor. The use of the IARC system might yet prove to be one of the few mechanisms that could reverse the privatization of biotechnology and challenge the direction of current research. If the IARCs are going to have any role at all in developing biotechnology for the needs of the rural poor, access to this type of information is very important. The real challenge for the IARCs responding to the biorevolution is to put into practice the lessons of the Green Revolution by appropriating biotechnology and gearing the research towards the specific problems of the rural poor.[10]

Aside from the international structures to facilitate the creation of indigenous research capacity, a crucial question remains as to whether developing countries will be able and committed enough to develop indigenous national research programs directed towards the needs of the rural and urban poor. Indonesia has already taken steps in this direction. However, the biotechnology programs still face serious technical, economic and political obstacles. The most important technical obstacles are the lack of trained personnel and the absence of a basic infrastructure. Additionally, long-term financial commitment to biotechnology is required.

THE INTEGRATED APPROACH

The mechanism implemented in the ABSP project, wherein a number of U.S. and Indonesian public and private institutions work together will also be implemented by the Indonesian government to enhance the application of biotechnology for agricultural development. These institutions share resources and expertise with counterpart institutions. An example of the integrated approach follows. The public institution from the U.S. Michigan State University, leads and coordinates the whole project. Private sector U.S. companies participating in the project are ICI Seed Company at Slater, Iowa and DNA Plant Technology in New Jersey. The participating public institutions from Indonesia are CRIFC and the Central Research Institute for Horticulture (CRIH) of the Agency for Agricultural Research and Development (AARD). In addition, the Directorate of Patents of the Ministry of Justice will also participate in the Intellectual Property Rights internship program. Fitotek Unggul is the only Indonesian private sector, company which will participate in the project.

The concept and spirit of resource sharing and the external funding provided by USAID are the fuel for the operation of this network. Effective and efficient communication mechanisms among the different member institutions are crucial to the success of the network. The ABSP project is not a usual biotechnology research and development project. Fundamental to its success is the accomplishment of the project goals through an integrated project-management approach which requires close collaboration among participating institutions.[11]

STRENGTHENING PERSONNEL CAPABILITY

The human-resource development dimension is one of the most important facets of the application of biotechnology. Research goals will be achieved in an inseparable relationship with the development of Indonesian scientists' knowledge of advanced biotechnology techniques used in both public- and private-sector institutions. Since biotechnology is new for Indonesian research systems, many of the personnel involved in agricultural biotechnology research have a very limited background in this area. They are breeders, pathologists, entomologists or microbiologists who may need additional training in various aspects of biotechnology. Further training in the most recent technologies is quite essential for strengthening personnel capability.

CHALLENGES AND ACCOMPLISHMENTS

Technology has driven changes in agriculture towards more effective and efficient production practices. The adoption of new technologies has improved the efficiency of agriculture production practices and resulted in current agricultural surpluses in many countries. Agricultural systems throughout the world continue to adopt new and better technologies that enable them to become more efficient and competitive in developing new markets and old markets for their agricultural products.

The inability of biotechnology to provide immediate help for major improvement of the needed traits in important Indonesian farm crops emphasizes the gap that exists between advances in molecular genetics and scientists' present-day knowledge of the biochemistry, physiology and genetics of important traits of crop plants. Little is known of the biochemical pathways leading from the gene to the final desired trait, nor of the physiology on a whole-plant basis of the biochemical pathways that affect plant development. Fortunately, biotechnology itself promises to be an excellent tool to help solve many of the questions that arise.

In early predictions of how rapidly the benefits of biotechnology might be expected, an important element of plant breeding and crop production was often overlooked. Once a series of new promising lines are developed to challenge those currently grown by farmers, extensive testing must be done. Only through appropriate testing of new promising lines in a wide range of environments, and exposure to different soils, weather, insects and diseases, can the truly superior ones be identified. However, any testing of biotechnology products should meet the regulatory requirements applied in the country.

Considering the urgent needs and available resources, development of agricultural biotechnology should be directed towards increasing productivity, stability, sustainability and equity of agricultural products. Furthermore, this technology should be used to supply raw materials for food, shelter, clothing and industry. Looking into the future, however, a manpower development program must be put in place that will ensure both a strong base and continuity. University courses may need to be enriched to better prepare scientists and technicians with a finer appreciation of biotechnology and greater flexibility to adapt to new challenges. It is important to realize that biotechnology is a highly dynamic field where a significant information gap can perceivably widen in only a few weeks or months. Scientists will need to be sent abroad for refresher courses, postdoctoral assignments, sabbatical leave and similar training to bring them up to par with their peers in developed countries.

It is important that adequate funds are available to ensure uninterrupted activity and high efficiency. An uncertain supply of funds is very detrimental to an ambitious program like biotechnology. Therefore, one recommendation is that the work plan should emphasize the allocation of necessary operational funds for an extended period of time.

INTELLECTUAL PROPERTY RIGHTS

Intellectual property rights (IPR) are traditionally defined to include the following six distinct types of protection patents, Plant Breeder's Rights, petty patents, trade secrets, copyrights and trademarks. The first four types are most commonly applied to living organisms. IPR, including patents for living organisms, is important for stimulating domestic private research and for facilitating access to inventions made externally. There are three major multinational agreements for the standardization and harmonization of patent laws. In addition, three regional conventions are operative which facilitate the application for and granting of multiple-country patents. One applies to the European Community countries, and one each to Anglophile and Francophile African countries.

Plant Breeders' Rights (PBR) is the common term applied to varietal protection such as promulgated under the International Convention for the Protection of New Varieties of Plants (UPOV). PBR, as the name suggests, is a patent-like system for protecting plant varieties.

Petty patents, also called utility models or simple patents, are in essence lesser patents for lesser inventions. They are typically allowed for adaptive inventions lacking an inventive step. The time period is typically shorter (e.g. 5 vs. 20 years) with grants registered rather than examined for merit.

Trade secrets are a distinct component of intellectual property laws which do not involve the granting of a certificate such as a patent. Rather, they are more general laws which allow a firm/individual to recover damages if something of commercial value is improperly acquired. Trade secrets would typically allow a maize breeding company to sue if its pure lines were stolen or if an employee revealed the crosses used. Trade secrets protection can also be applicable to databases and customer lists, and in fact anything of commercial value that a firm attempts to keep secret. Trade secrets often are used in conjunction with patents or other forms of IPR.

The general situation with living organisms is the same as for any other form of invention, with two key distinctions. First, advancements in this area have been rapid, so that the inventor has a limited time to recover the investment. The private investor is then concerned for strong protection across a broad market, often covering multiple countries. Second and most important, many protected living organisms such as seeds have the ability to regenerate themselves. This means that secrecy cannot be used, and legal protection is the only practical form of protection available. Hence, private biotech firms are very concerned with IPR. Despite the significance of IPR as an economic tool, there is little direct evidence that they do indeed increase investment in research and development (R&D). The evidence is particularly limited for developing countries. In addition no one can demonstrate that the details of the laws, the key choices over duration, scope, etc., are optimal. That said, what evidence exists supports the theoretical expectation that IPR does indeed enhance private R&D.

The situation with patents in developing countries is somewhat different as the case can be made that these countries do not compete in the forefront of technological development. Under those circumstances, patents are referred to as "protecting import monopolies." Even in the event that a country cannot contribute at the cutting-edge of technology, two additional factors need to be considered. First, patents leading to local licensing of imported technologies and products do lead to domestic, adaptive invention. In the absence of the imports, the added employment and skill improvement from adaptation would be lost. Second, adequate IPR are important in accessing products and technologies made elsewhere. Without IPR, owners of inventions, fearful of losing those inventions and encountering competition in developing-country markets, will limit and restrict access to those technologies and products. This can mean a national economy is restricted to old technology which may be uncompetitive in the integrated world market. Indonesia, as a likely major adopter of agro-biotechnology products for the midterm future, needs to be conscientious of its access to these products.

The role of IARCs and other organizations such as the International Service for the Acquisition of Agri-Biotech Applications (ISAAA) and the Intermediary Biotechnology Service (IBS) should be to function as intermediaries between technology developers and their users.

BIOSAFETY PROCEDURES

Indonesia currently lacks a biosafety program. The recently-established Committee on Biosafety of the National Committee on Biotechnology will be recommending that the U.S. National Institutes of Health laboratory protocols be adopted. Consideration has not yet begun on the field-testing of genetically engineered organisms, and its lack could soon delay access to important biotech developments. ICI Seed Co., for example, plans to test a *Bacillus thuringiensis*-augmented corn variety for Asia stemborer resistance in Indonesia in 1995 through the ABSP project. Identification and adoption of the appropriate regulations should be given the highest priority. The following steps are recommended:

- Identify pertinent risks and probable effects and balance with possible beneficial outcomes, e.g. reduced pesticide use.
- Because risks for plants are less than for microorganisms, plants should be addressed first, microorganisms and animals subsequently.
- In the interests of time and international harmonization, existing laws worldwide should be examined for applicability and suitable ones adopted or adapted as appropriate. There are pragmatic reasons to consider U.S. law, including experience and requirements under the AID/MSU ABSP project.
- Regulations should be harmonized within ASEAN, using plant quarantine agreements as a pattern. In this regard ISAAA organized biosafety workshop among ASEAN countries which was held in Indonesia in 1993.

REFERENCES

1. Evans DA, Sharp WR, Ammirato PV, Yamada Y. Handbook of Plant Cell Culture. Vol. 1. Techniques for Propagation and Breeding. New York: Macmillan Publishing Co, 1983.
2. Vasil IK. Cell Culture and Somatic Cell Genetics of Plants. Vol. 6. Molecular Biology of Plant Nuclear Genes. New York: Academic Press, 1989.
3. Mariska I, Sukmadjaja D, dan Gati E. Perkembangan penelitian bioteknologi kultur jaringan tanaman obat. Edisi khusus Penelitian Tanaman Rempah dan Obat 1989; 5(2):8-18. (Special edition on research progress in tissue culture of medicinal plants).
4. Grierson D, Covey SN. Plant Molecular Biology. 2nd Edition. New York: Blackie and Son, 1988.
5. CBS. Statistical year book of Indonesia 1988. Jakarta: CBS, 1989.
6. Manwan SI, Blumenschein A. National Coordinated Research Program: Corn. Central Research Institute for Food Crops. Bogor: CRIFC, 1988.
7. Dimyati A, Manwan I. National Coordinated Research Program: Cassava and Sweet Potato. Bogor: Central Research Institute for Food Crops, 1992.
8. Persley GJ. Beyond Mendel's Garden: Biotechnology in the Service of World Agriculture. Wallingford: CAB International, 1990.
9. Hobbelink H. Hope or False Promise? Biotechnology and Third World Agriculture. New York: The International Coalition for Development Action, 1987.

10. Sasson A. Biotechnology and Development. UNESCO. Paris: Technical Centre for Agricultural and Rural Cooperation (CTA), 1988.
11. Bedford B. What ABSP will accomplish in Indonesia. Biolink 1993; 1(2):3-4.
12. CBS. Statistical year book of Indonesia 1987. Jakarta: CBS, 1988.
13. CBS. Statistical year book of Indonesia 1989. Jakarta: CBS, 1990.
14. CBS. Statistical year book of Indonesia 1990. Jakarta: CBS, 1991.

CHAPTER 8

Present Status of Plant Biotechnology Research and Development in Malaysia

F.C. Low, M.D. Hassan, S.C. Cheah, M.Y. Aziah,
K.F. Cheong, J. Hafsah, R. Wickneswari, B. Krishnapillay

INTRODUCTION

In recent years, Malaysia has recorded impressive growth, especially in the industrial sector. In spite of this rapid industrial development, crop production remains as one of the important components of the nation's economy. Apart from food production, the agricultural sector is expected to grow and assume the role of being a supplier of raw materials for the manufacturing industries. Agriculture is traditionally linked to the rural sector, which is generally poor. Thus, improvement in the livelihood of the rural sector is a firm commitment of the government. The application of science and technology (S&T) is envisaged as a vehicle to modernize the agricultural sector. Malaysia aspires to be a developed nation by the year 2020. In order to achieve that status, rapid transformation in the agricultural sector will be required. Biotechnology has been identified as a key component in the revolution of crop production in the country.

Biotechnology can be defined broadly as the economic application of biological processes and enhanced science for the benefit of mankind. In the context of modernizing the agricultural sector, biotechnology aims to promote economic sustainability and growth of the sector. These are achieved through:

Plant Biotechnology Transfer to Developing Countries,
edited by D.W. Altman and K.N. Watanabe. © 1995 R.G. Landes Company.

- increasing productivity
- reducing production costs
- minimizing risks
- enhancing returns
- facilitating the adoption of new technologies
- improving the quality of produce/products, to meet consumer and market demands.

Rapid advances achieved in the application of biotechnology elsewhere have prompted Malaysia to earnestly embark on establishing biotechnology research and development (R&D) to improve crop productivity. In 1990 and 1991, about R.M. 9.7 million (U.S. $4 million) and R.M. 19.62 million (U.S. $8 million) respectively, were allocated by the government to strategic research, some of which included plant biotechnology. This fund was channeled to R&D institutions and universities in the country through a mechanism called the Intensification of Research in Priority Areas (IRPA). A considerable number of biotechnology research projects have been initiated in the country as a consequence of this support. This chapter reports on the current status of research and achievements in plant biotechnology R&D in Malaysia.

RESEARCH INSTITUTIONS AND ORGANIZATIONS

Malaysian agriculture is primarily based on high-volume commodity crops like rubber, palm oil, cocoa and timber. Research institutions were established according to the needs and importance of those commodities in the economy of the country. Hence, research institutions such as the Rubber Research Institute of Malaysia (RRIM), Palm Oil Research Institute of Malaysia (PORIM) and Malaysian Cocoa Board (MCB) were established specifically to conduct research in rubber, oil palm and cocoa, respectively. Of these three, the RRIM is the oldest (about 75 years), MCB is the youngest (less than five years) and PORIM is in-between, at about 10 years. The Forest Research Institute of Malaysia (FRIM), similarly, executes research on forest tree species. Plant biotechnology R&D is carried out mainly in these and other public institutions. All the above research institutions are administered under the Ministry of Primary Industries of Malaysia.

The Malaysian Agricultural Research and Development Institute (MARDI) carries out research on crops other than the major ones described above. These include rice, cassava, fruits, other food crops, tobacco and ornamental plants. The institute is placed under the Ministry of Agriculture.

In addition to these major research institutions, applied plant biotechnology is also carried out at the local universities. However, these universities, namely the University of Malaya, the National University of Malaysia, the Science University of Malaysia and the Agricultural University of Malaysia are focused mainly on basic research and under the preview of the Ministry of Education.

BIOTECHNOLOGY OF INDUSTRIAL CROPS

RUBBER

A two-prong approach in plant biotechnology studies in rubber (*Hevea brasiliensis*) has been adopted in the RRIM. The two areas selected for study are in vitro culture and molecular biology.

Hevea in vitro technology has been established in the RRIM for nearly 30 years. Current efforts are aimed at protocol development for somatic embryogenesis of elite clones for transgenic and mutant plant production. Recent studies have indicated that the origin of *Hevea* embryoids via anther and ovule culture is a single cell.[1] Preliminary investigation on embryo rescue has yielded some success and further emphasis on this project is envisaged. In the next phase of *Hevea* in vitro R&D, studies will be intensified towards the development of improved techniques for regeneration and propagation of products of plant transformation. Investigation to determine whether the current trend of minimizing plantlet recovery via in vitro culture and the use of special target cells for plant transformation are applicable to *Hevea* will be carried out.

Studies on the molecular biology of *Hevea* are focused on three aspects. These are DNA polymorphisms, chloroplast genome and plant transformation. DNA-marker techniques such as RFLPs and RAPDs have been established for *Hevea*. With the use of RFLPs, DNA polymorphisms between species as well as within species have been demonstrated.[2] Additionally, interspecific polymorphism between two clones which share a common female parent was reported.[3] More recently, marker techniques like sequence-tagged sites (STS) and microsatellites (MS) have also been successfully applied to *Hevea* studies.[4] DNA polymorphisms in sequence-tagged microsatellite sites (STMS) have been demonstrated between *Hevea* species (Fig. 8.1) as well as within the *H. brasiliensis* species (Fig. 8.2).[4] Work is in progress on the applications of these DNA-marker techniques for *Hevea* parental and cultivar identification, genetic relatedness in a new germplasm collection, genetic homogeneity in *Hevea* plants derived from in vitro culture and the construction of a genetic linkage map. Work on DNA fingerprinting of rubber clones and plants from in vitro culture with tandem repeat DNA sequences is also in progress. Libraries containing genomic and cDNA from leaves as well as latex have been constructed.

For *Hevea* chloroplast genome studies, chloroplast DNA (ctDNA) libraries in pUC19 plasmid and phage Lambda vectors have been constructed. The gene (*rbcL*) for the large subunit (LS) of the photosynthetic enzyme, D-ribulose-1,5-bisphosphate carboxylase/oxygenase (RUBISCO) have been isolated, sequenced and characterized.[5] The availability of *Hevea*'s *rbcL* nucleotide sequence has made it possible to deduce the amino acid sequence of its LS protein and study its active sites. Maternal inheritance of cytoplasm based on ct DNA-restriction endonuclease profile pattern stud-

 Plant Biotechnology Transfer to Developing Countries

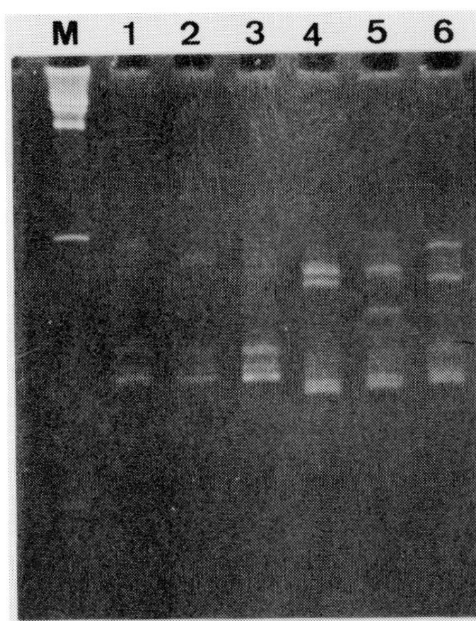

Fig. 8.1. Polymorphism in sequence-tagged microsatellite (GA)$_9$ site seen between several Hevea *species. Lanes 1-2,* H. pauciflora*; Lane 3,* H. guianensis*; Lane 4,* H. camargoana*; Lane 5,* H. benthamiana*; Lane 6,* H. brasiliensis*; M1, M. wt. marker, kb ladder.*

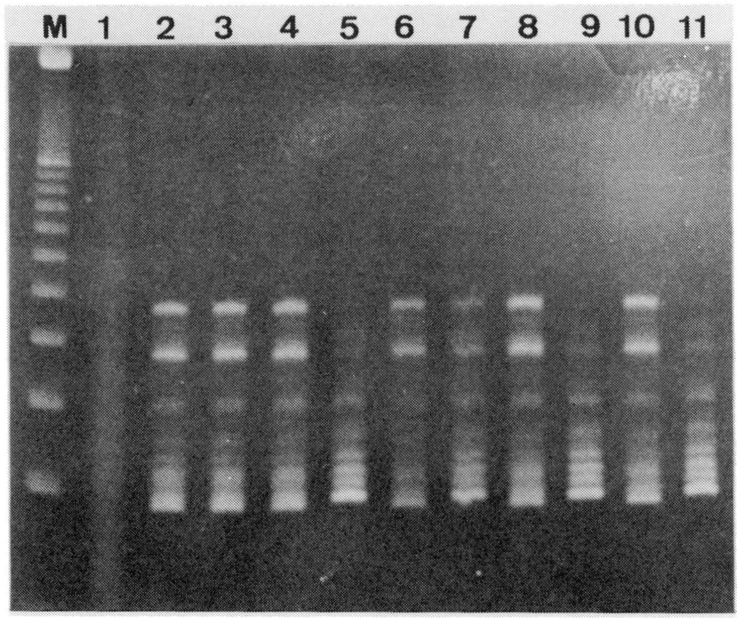

Fig. 8.2. Polymorphism in sequence-tagged microsatellite (GA)$_9$ site seen between several H. brasiliensis clones, after vertical electrophoresis through 6% polyacrylamide gel; Lane 1, Control; Lane 2, Tjir 1; Lane 3, PB 86; Lane 4, RRIM 600; Lane 5, PB 5/51; Lane 6, PB 49; Lane 7, PB 260; Lane 8, RRIM 713; Lane 9, RRIM 717; Lane 10, PB 235; Lane 11, PB 265; M, M.wt. marker, 100 bp ladder.

Fig. 8.3. Detection of RFLP in oil palm (Elaeis sp.). DNA obtained from various oil palm varieties were digested with the restriction enzyme DraI and hybridized with cDNA probe, pOP-G240. The varieties albescent (alb), virescent (vir), dura (D), pisifera (P), Mantled (M) and dumpy (Dy) belong to the species E. guineensis. *The* E. oleifera *(Eo) sample shown here is from a palm of South American origin.*

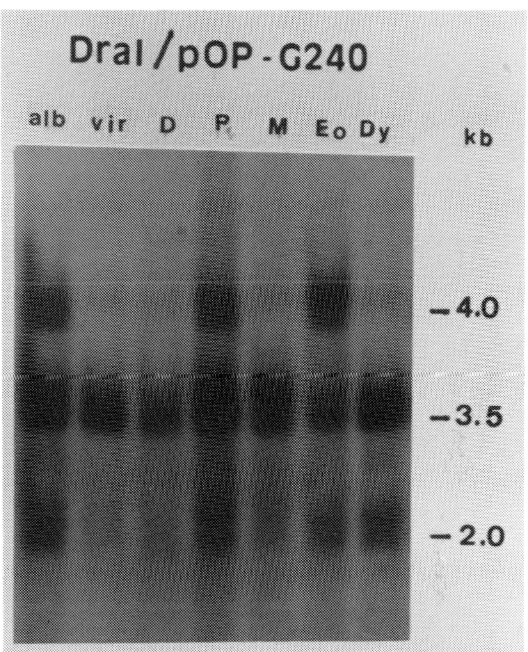

ies has been observed for some of the *H. brasiliensis* clones. Phylogenetic studies of *Hevea* in relation to other plant species through comparisons of their *rbcL* gene sequences are currently being pursued.

Transformation of *Hevea* tissues with reporter genes, e.g. GUS, has been initiated. DNA delivery systems which have been investigated include DNA imbibition, co-cultivation with *Agrobacterium tumefaciens* and particle bombardment.[6] Preliminary results are encouraging and appear to indicate stable transformation, suggesting that *Hevea* transformation by a foreign gene is possible. Present research activity is directed towards the transformation of *Hevea* by useful foreign genes of importance and the use of tissue-specific promoters to target these genes for expression in the appropriate tissue of the tree.

OIL PALM

PORIM currently works on biotechnology research projects aimed at fulfilling both the short- and long-term needs of the industry. The areas of research include oil quality, tissue culture and genetic studies. Most of the projects are directed towards the development of tools and techniques required for genetic engineering of oil palm for the production of oil high in mono-unsaturates. It is envisaged that such an oil will facilitate the entry of palm oil into the liquid oil market as well as provide oleic acid feedstock for the oleochemical industry.

In order to achieve the above objective, a particular strategy is employed. Research is focused on the manipulation of genes of the enzymes controlling the synthesis of the mono-unsaturated oleic acid in the oil palm mesocarp, such that palmitate is diverted to form oleate in the lipid biosynthetic pathway. Currently, efforts towards the purification and characterization of enzymes of interest, as a preliminary step to the isolation of their genes are being carried out. Studies also are being conducted on the temporal and tissue-specific DNA sequences which control fatty acid biosynthesis in the oil palm mesocarp.[7] It is envisaged that these sequences will be useful for directing the expression of genes of interest in transgenic palms. As in all attempts at genetic engineering, the availability of a stable transformation technique is a prerequisite. Attempts are being made to achieve this by various means, namely through protoplast, pollen, desiccated embryo and cell culture using the techniques of DNA uptake, electroporation and biolistics.

Another aspect of oil quality is its free fatty acid content. Work has been initiated at PORIM to study the process of lipolysis after the fruits have been harvested. An endogenous lipase has been detected in the mesocarp[8] and current research is aimed at purifying the lipase enzyme.

RFLP, RAPD and DNA fingerprinting probes have been applied to assess genetic variability in the oil palm (Fig. 8.3). Results showed about 30% variability exists between the species *Elaeis guineensis* (the West African oil palm) and *E. oleifera* (the South American oil palm).[9] However, variability within the *guineensis* species is only about 7%. RFLP probes were also used to detect changes in palms derived from tissue culture.[10] A quality control process for oil palm tissue culture has been developed using a set of hypervariable RFLP probes. Efforts are underway to construct genetic maps.

DNA fingerprinting of palms using simple repetitive DNA probes $(GATA)_4$ and $(GACA)_4$ have indicated that such probes are able to distinguish palm species. These probes are therefore useful for ecological and phylogenetic studies of palms.

As the economic product of the oil palm is the fruit, the flowering process has great effect on yield. Studies on this developmental process are directed towards the isolation of gene sequences which are specifically-expressed during flowering in the oil palm. Two-dimensional PAGE, differential hybridization and subtractive hybridization are some of the techniques used in the studies.

The process of micropropagation of oil palm has reached an advanced stage of development. Reports in the mid-1980s indicating that some clones developed abnormally during flowering were a setback in its commercialization. Research aimed at understanding the basis of this abnormality was therefore intensified.

Forest Trees

The forestry sector is embarking on the establishment of forest species plantation to overcome the shortage of timber and nontimber resources. In most species, seeds are difficult to procure and the unavailability of planting materials will impede the success of the plantation program. Hence, vegetative propagation techniques are being developed to overcome the shortage of planting materials and to provide a rapid means of propagating superior genotypes.

Micropropagation techniques at FRIM are being developed for nontimber species (rattan and bamboo), timber species (Dipterocarps and nondipterocarp species) and exotic species. Two species of rattan, *Calamus manan* and *C. caesius* have been micropropagated.[11] In both these species, the collar region of aseptically-germinated seedlings are excised and plantlets are regenerated via the shoot proliferation method. Regeneration of *C. manan* plantlets is also possible through somatic embryogenesis in callus originating from embryos[12] and young leaf tissue of mature plants.[13] Field testing of tissue culture-raised seedlings via the shoot proliferation method is being conducted.[14]

Three species of bamboo, *Bambusa vulgaris, B. vulgaris* var *strata* and *Gigantochloa levis* have been micropropagated successfully.[15] An exotic species, *Acacia mangium* has also been micropropagated successfully and is currently undergoing field testing.[16]

Successful micropropagation techniques have been developed for three indigenous forest timber species, namely *Dyera costulata* (or Jelutong by its local name), *Endospermum malaccense* and *Gonystylus banccanus*. All three species have been selected for cultivation as timber forest plantation. *Dyera costulata* and *Endospermum malaccense* will be cultivated under normal soil, while *Gonystylus banccanus* has been identified as the only species suitable for peat-form areas.

Cryopreservation techniques are being developed for recalcitrant-seeded tropical forest species for conservation of genetic diversity as well as sustained and improved utilization of these species.[17] To date, cryopreservation protocols have been developed for four bamboo species (*Thyrsostachys siamensis, Bambusa arundinacea, Dendrocalamus brandisii* and *Dendrocalamus membranaceus*),[18] one rattan species (*Calamus manan*),[19] one nondipterocarp species (*Sweitenia macrophylla*) and two dipterocarp species (*Dipterocarpus alatus* and *Dipterocarpus intricatus*).[20]

Information on the genetics of Malaysian forest plant species is extremely limited. Such information is essential for successful tree improvement and conservation programs. Protocols for genetic screening based on molecular markers such as isozymes, RFLPs and RAPDs have been developed for the evaluation of genetic diversity, gene-dispersal patterns and phylogenetic studies. The genetic information thus generated will provide guidelines for:

- delimitation of natural populations to be preserved, e.g. virgin-jungle reserves and national parks;
- sampling strategies for ex situ conservation and tree improvement programs;
- management of stands for natural regeneration;

- assessment of inbreeding effects and loss of genetic variation resulting in erosion of viability of small and fragmented plant populations;
- tracking desirable genes for tree improvement, in species selected for plantation programs.

Techniques for isozymes analysis after starch gel electrophoresis have been developed for *Acacias*.[21] Using these techniques, the genetic diversity and breeding system parameters of several species, including *Acacia auriculiformis*[22,23] *Acacia mangium* (Wickneswari, unpublished), *Anisoptera laevis, Dipterocarpus oblongifolius, Dryobalanops aromatica, Dryobalanops oblongifolia, Hopea nervosa, Hopea odorata, Neobalanocarpus helmii* and *Shorea acuminata*[24-26] have been determined. Positive and early identification (i.e. at the seed and 1-2 week-old seedling stage) of spontaneous and manipulated *Acacia mangium* x *Acacia auriculiformis* hybrids have been carried out using the same isozyme technique.[27-29]

Protocols for total DNA extraction of about 30 dipterocarp tree species and about 10 nondipterocarp tree species including rattan and bamboo species for RFLP and RAPD analysis have been developed (Wickneswari, unpublished). The phylogeny of *Dipterocarpaceae* has been constructed using RFLP analysis of PCR-amplified chloroplast genes.[30,31] Methods for analysis of mitochondrial gene polymorphisms in *Dryobalanops aromatica* and *Hopea odorata* are being developed so that genetic diversity can be determined using organellar genes.[26]

Cocoa

Studies on the molecular biology and in vitro culture of cocoa were initiated only recently at the Malaysian Cocoa Board (MCB), it being a newly-established research organization.

BIOTECHNOLOGY OF FOOD CROPS AND ORNAMENTAL PLANTS

Rice

Rice, the strategic crop of Malaysia, faces many challenges from a wide range of yield limitations. Some of these challenges are better addressed through conventional approaches, while others are more appropriately dealt with through biotechnological approaches. One of the major limitations in rice production is pests and diseases. As much as 20-35% of the total yield, at an estimated cost of R.M. 150-250 million (U.S. $60-100 million), has been lost annually owing to pest and diseases. Among the several diseases of rice, the rice tungro disease, which is locally known as *penyakit merah*, is the most important. This disease is caused by a complex of two viruses, namely rice tungro bacilliform virus (RTBV) and rice tungro spherical virus (RTSV). Lack of naturally-occurring resistance genes has hampered efforts to breed for resistance against the tungro virus in rice. However, the introduction of resistance through viral coat protein-mediated resistance promises to be useful. A viral coat protein gene ligated to an appropriate construct is introduced into the plant by this procedure and allowed to be expressed. Expression of the viral coat-protein gene in plant cells has been shown to provide protection against the virus.[32]

A project to develop transgenic rice plants conferring resistance against tungro virus infection using the coat protein-mediated resistance has been initiated. The work is carried out in collaboration with the International Laboratory for Tropical Agricultural Biotechnology (ILTAB)/The Scripps Research Institute (TSRI), La Jolla, California, U.S. under the International Program on Rice Biotechnology of the Rockefeller Foundation.

The coat-protein gene for RTBV from the Philippines has been cloned and sequenced by Qu et al.[33] Transgenic plants raised from various gene constructs using the cloned coat protein-gene were screened for expression of the RTBV coat-protein gene.[34] R_1 seeds from positive plants were then sent to MARDI for screening against the Malaysian isolate of RTBV. In the screening procedure, plants were inoculated three times with rice tungro virus complex to reduce the probability of false positives. Initial results have been encouraging. If this approach is shown to be useful, Malaysian rice varieties which are presently released for cultivation will be transformed with the viral coat-protein gene.

Several gene delivery systems to introduce desired genes into rice have been evaluated. Of these, the particle bombardment method is found to be the most promising. At this stage, the β-D-glucuronidase (GUS) reporter gene is used in the transformation experiments to optimize delivery conditions.

Histochemical assays of GUS activity on the bombarded tissues showed distinct blue spots, indicating potentially positive results of transformation and expression of the introduced gene. Rice plants have been regenerated from the bombarded tissues. Histochemical assay of GUS as well as molecular screening of the regenerated plants have confirmed transgenic expression.

PAPAYA

The emergence of new diseases and pests in the country has affected crop production significantly. Papaya ringspot virus was unknown in Malaysia until a large outbreak in the state of Johor was reported two years ago. This virus has adversely affected the papaya industry in the country. The popular *Eksotika* papaya variety has proven to be susceptible to this virus, and resistant gene material is unavailable in the gene pool. A project was initiated recently at MARDI to use the coat protein-mediated resistance approach. The coat-protein gene from the virus was isolated and utilized for development of the gene construct to be transferred into papaya. A cDNA library was subjected to amplification of the coat-protein gene of the virus by polymerase chain reaction (PCR). A DNA fragment of 800 bp was isolated, cloned and sequenced. Currently, the cloned fragment is being used to develop a gene construct. These gene constructs will be transferred into papaya when a transformation system for papaya is established.

ORNAMENTALS AND OTHERS

At MARDI, tissue culture research in other crops was initiated much earlier than molecular biology research. Various crops such as banana, cassava, orchid, pineapple, muskmelon, strawberry, potato, sweet potato and fragrant-flowered species have been successfully propagated through tissue culture. Significant among the achievements in this area of research is the development of a successful in vitro propagation protocol for cocoa.

INTERNATIONAL COLLABORATION

Several international agencies were instrumental in helping Malaysia launch her plant biotechnology program. The Rockefeller Foundation, under its International Program on Rice Biotechnology and in collaboration with the International Laboratory for Tropical Agricultural Biotechnology (ILTAB)/The Scripps Research Institute (TSRI), La Jolla, California (U.S.), assisted Malaysia in developing transgenic rice plants enhanced by resistance against the tungro virus. Similarly, RFLPs and other DNA-marker techniques for *Hevea*, were developed under the Rockefeller Foundation career fellowship for biotechnology, in collaboration with the Cambridge Laboratory (United Kingdom). The development of these techniques resulted in their rapid applications to *Hevea* studies.

Crop improvement through mutation breeding in Malaysia was greatly assisted by participation in training courses conducted by the Food and Agriculture Organization/International Atomic Energy Agency (FAO/IAEA) in Seibersdorf, Austria. Courses such as the "FAO/IAEA Inter-regional Training Course on the Induction and Use of Mutation in Plant Breeding" have greatly accelerated progress in mutation breeding in several crops in the country, such as banana, rice and *Hevea*. Additionally, the development of fingerprinting techniques for in vitro culture plants of *Hevea* was facilitated by a research contract from the same agency.

Programs in forestry, particularly in conservation and biodiversity characterization were carried out separately with several international collaborations. The development of micropropagation techniques for rattan was carried out with an initiative from the International Development Research Center (IDRC) of Canada. The Australian Center for International Agricultural Research (ACIAR) contributed over a period of four years to the project on the development of isoenzyme analysis of *Acacia*. Studies on the genetic diversity and breeding systems of *Dipterocarpaceae* were conducted as part of a joint research project between the National Institute of Environmental Science (NIES) Japan, the Forest Research Institute of Malaysia (FRIM) and the University of Agriculture of Malaysia (UPM). This six-year collaboration was initiated in 1990. The evaluation of genetic diversity and characterization of breeding systems of some important rainforest plant species were carried out as a sub-project under the overall program of biodiversity and conservation, under the sponsorship of the Overseas Development Administration (ODA) of Britain. Collaboration from the International Plant Genetic Resources Institute (IPGRI) on the development of cryopreservation methods for recalcitrant-seeded forest species began in early 1994 and is expected to conclude at the end of the same year.

In 1991, a British High Commissioner's Research Attachment Award for three months at the University of Hertfordshire (formerly, Hatfield Polytechnic) in the United Kingdom provided the impetus for the transformation of *Hevea*. The biolistic gun at the Rothamstead Research Station was used during that period for *Hevea* genetic transformation by particle bombardment. Through the assistance of the International Service for the Acquisition of Agribiotech Applications (ISAAA), an attachment of six months at the Institut fur Genbiologische Forschung Berlin (IGF) Germany was made possible by financial support from Schering AG. This helped progress in vector construction and provided the next phase of development in genetic transformation studies of *Hevea*.

CONCLUSION

The rapid progress in plant biotechnology in Malaysia is testimony to the readiness of the country to learn and work with others and to employ high technology in the modernization of its agriculture. The applications of molecular biology and in vitro techniques facilitated through collaborations with various international agencies, coupled with the receptivity of her researchers to innovate and to adapt to new and improved biotechnologies have contributed significantly to increased agricultural productivity in Malaysia. With continuous support from various international agencies and the government, the future of plant biotechnology in Malaysia is envisaged to be bright, robust and dynamic.

REFERENCES

1. Samsidar H, Hafsah J. Anatomical studies of embryoid development from somatic callus of *Hevea brasiliensis*. Submitted to Physiologia Plantarum 1994.
2. Low FC, Gale MD. Development of molecular markers for *Hevea*. J Nat Rubb Res 1991; 6:152-157.
3. Low FC, Siti Arija MA, Chow KS et al. Restriction fragment length polymorphisms as probes to plant diversity. Conservation of Plant Genetic Resources Through *In Vitro* Methods ISBN 967-99915-2-0, 1991:110-117.
4. Low FC, Safiah A, Khoo SL et al. The application of some recent DNA fingerprinting techniques-Microsatellites and sequence-tagged sites for *Hevea* studies. Proc Second Symp on Trends in Biotech 1994; 122-124.

5. Cheong KF, Koh CL. Nucleotide sequence of the chloroplast gene for the large subunit of ribulose-1,5-bisphosphate carboxylase/oxygenase from *Hevea brasiliensis*. J Nat Rubb Res 1993; 8:146-153.

6. Cheong KF, Arokiaraj P, Yeang HY. Genetic transformation of the rubber tree, *Hevea brasiliensis*. Proc Fourth Seminar on National Biotech Programmes 1992;18-19.

7. Siti Nor AA, Farida HS, Cheah SC. Detection of differentially expressed genes in the development of oil palm mesocarp. AsPac J Mol Biol Biotechnol 1994; 2:113-118.

8. Sambanthamurthi R, Chong CL, Oo KC et al. Chilling-induced lipid hydrolysis in the oil palm (*Elaeis guineensis*) mesocarp. J Exp Bot 1991; 42:1199-1205.

9. Cheah SC, Siti Nor AA, Ooi LCL et al. Detection of DNA variability in the oil palm using RFLP probes. Proc 1991 PORIM International Palm Oil Conf Agric (Module 1), 1993:144-150.

10. Cheah SC, Wooi KC. Application of molecular marker techniques in oil palm tissue culture. Proc ISOPB International Symp on "Recent Developments in Oil Palm Tissue Culture and Biotechnology", Kuala Lumpur, Malaysia, 1993.

11. Aziah MY. Tissue culture of rattans. In: Wan Razali WM, eds. A Guide to the Cultivation of Rattan. Malayan Forest Record. 35. 1992:149-161.

12. Aziah MY. Shoot formation of *C. manan* under in vitro. Proc of Seminar on Tissue Culture of Forest Species. In: Rao AN, Aziah MY, eds. FRIM/IDRC, Kepong, 1989:45-49.

13. Aziah MY, Darus HA. Potentials of in vitro propagation techniques in Malaysian Forestry. Paper presented at the International Symp on Application of Plant in vitro Technology. Serdang, Malaysia, 1993.

14. Aziah MY, Nur Supardi MN. Initial growth of tissue culture raised *C. manan* seedlings. RIC bulletin 1993; 12:6-7.

15. Aziah MY, Darus HA. Production of bamboo propagation via tissue culture for large scale planting programme. First National Bamboo Seminar. Forest Research Institute Malaysia, Kulal Lumpur, Malaysia, 1992

16. Darus HA. Rapid propagation of *Acacia mangium* plantlets using micropropagation techniques. In: Sheikh AA et al, eds. Recent Developments in Tree Plantation of Humid/Sub Humid Tropics of Asia, 1989:202-216.

17. Krishnapillay B, Marzalina M. Preservation of some tropical forest tree seeds. Proc Fifth National Biotechnology Seminar, Port Dickson, 1993.

18. Krishnapillay B, Marzalina M, Haris M. Liquid nitrogen storage prolongs bamboo seed longevity. Buletin Buluh 1993; 2:1-3.

19. Krishnapillay B, Marzalina M, Aziah MY et al. Cryopreservation of excised embryos of rattan manan (*Calamus manan*). Proc Third National Biotechnology Seminar, Kuantan, 1991.

20. Krishnapillay B, Marzalina M, Paki Hayacamee P et al. Cryopreservation of *Dipterocarpus alatus* and *Dipterocarpus intircatus* for long-term storage. 23rd. International Seed Testing Congress (ISTA). Buenos Aires, Argentina, 1992.

21. Wickneswari R, Norwati M. Techniques for starch gel electrophoresis of enzymes from *Acacias*. In: Breeding Technologies for Tropical Acacias. ACIAR Proc 37, Canberra: ACIAR, 1992:88-100.

22. Wickneswari R, Norwati M. Spatial heterogeneity of outcrossing rates in *Acacia auriculiformis* in Australia and Papua New Guinea. International Sym on Population Genetics and Gene Conservation of Forest Trees. Bordeaux, France, 1992.

23. Wickneswari R, Norwati M. Genetic diversity of natural populations of *Acacia auriculiformis*. Australian J Bot, 1993; 41:65-77.

24. Wickneswari R. Breeding systems and genetic diversity in *Dryobalanops aromatica* Gaertn f. International Symposium on Population Genetics and Gene Conservation of Forest Trees. Corvallis, Oregon, USA, 1990.

25. Norwati M, Wickneswari R, Zakri AH. Jelutong-a lowland rainforest tree species of high genetic diversity. Poster paper presented at Conf. on Forestry and Forest Products Research, FRIM, Kuala Lumpur, 1993.

26. Wickneswari R, Norwati M, Zawawi I. Measuring genetic diversity of *Dipterocarpaceae* using molecular markers. Seminar on Measuring and Monitoring Biodiversity in Tropical and Temperate Forests, Chiang Mai, Thailand, 1994.

27. Wickneswari R. Use of isozyme analysis in a proposed *Acacia mangium* x *Acacia auriculiformis* hybrid seed production orchard. J Tropical Forest Sci 1989; 2:157-164.

28. Wickneswari R, Norwati M. Pod production and hybrid seed yield of *Acacia mangium* and *Acacia auriculiformis*. In: Breeding Technologies for Tropical Acacias. ACIAR Proc 37, Canberra: ACIAR, 1992:57-62.

29. Sesgley M, Harbard J, Smith RM et al. Reproductive biology and interspecific hybridisation of *Acacia mangium* and *Acacia auriculiformis* A. Cunn. ex Benth. (Leguminosae:Mimosideae). Australian J Bot 1992; 40:37-48.

30. Tsumura Y, Wickneswari R. Chloroplast DNA variation in *Dipterocarpaceae*, Preliminary results. 5th National Molecular Biology Seminar, Kuala Lumpur, 1992.

31. Tsumura Y, Wickneswari R, Kawahara T et al. Phylogeny of *Dipterocarpaceae* using RFLP of chloroplast DNA. XVth International Botanical Congress, Yokohama, Japan, 1993.

32. Powell AP, Nelson RS, Hoffmann DeB et al. Delay of disease development in transgenic plants that express the tobacco mosaic virus coat protein gene. Science 1986; 232:738-743.

33. Qu R, Bhattacharyya M, Laco GS et al. Characterization of the genome of rice tungro bacilliform virus: comparison with commelina yellow mottle virus and caulimoviruses. Virology 1991; 185:354-364.

34. Fauquet CM, Beachy RN. Engineering rice towards tungro virus resistance and rice yellow mottle virus resistance. Proc Sixth Ann Meeting of the International Program on Rice Biotech Feb. 1-5, 1993. Chiang Mai, Thailand, 1993:10.

CHAPTER 9

STATUS OF DEVELOPMENT IN PLANT BIOTECHNOLOGY IN THAILAND

Sutat Sriwatanapongse and Sakarindr Bhumiratana

INTRODUCTION

Biotechnology-based industry is very important to the Thai economy. Being predominantly an agro-industrial country, the Kingdom boasts high current outputs with great potential for expansion in food and other agro-based raw materials production for which biotechnology can be utilized for cost reduction, quality improvement and added value by further conversion to other products. Thailand is currently the world leader in the export of rice, cassava products, canned pineapple and canned baby corn. It is quite clear that Thailand will remain agricultural-based, with more than 50% of the population predicted to be in the agricultural sector in the beginning of the 21st century. With the growth of the economy, new technology will continually be introduced to replace the more conventional, more traditional production practices, and export items will be diversified at a quickening pace. The food processing industry is currently the largest importer of foreign technology, in terms of expenditure on technology from abroad classified by product type. In terms of employment, the food processing industry is running neck and neck with the country's other principal industry and top employer, textiles.

Recognizing the potentials of biotechnology to affect a broad spectrum of industries, Thailand has placed increased emphasis on biotechnology over the last decade, with the establishment of the National Center for Genetic Engineering and Biotechnology (NCGEB) in 1983, and the launching of the Science and Technology Development

Project in 1985. These two activities merged as part of the National Science and Technology Development Agency (NSTDA) at the end of 1991. With this push, it is hoped that biotechnology will play a significant role in the continued growth of the Thai economy in the year 2000 and beyond. In particular, new biotechnology will be transferred more efficiently and local R&D activities will achieve adequate progress and produce more and more concrete outputs for commercialization and public utilization. The intention of this chapter is to give an overview on the current status of development in plant biotechnology in Thailand.

CURRENT TECHNOLOGICAL CAPACITY

The level of technological capability in the area of biotechnology has improved considerably during the past decade, largely due to the resources made available by the government and international agencies. Between 1982-1989 more than 700 million bahts worth of funding from five major sources were granted. This figure does not include numerous other smaller grants provided by the National Research Council and the internal funding of various universities and research institutions. Of the major funding sources, STDB (Science and Technology Development Board—The Thai-US Cooperation Project) and NCGEB combined for just over half of the total amount with appreciable growth, while the other three sources have declined. The total available funding is expected to grow continuously for the next decade. The resulting R&D activities and graduate scholarship programs provided by NCGEB and STDB are the major contributors in building up the technological capacity of the country. This capacity resides mainly in various laboratories of the major universities.

A recent study carried out by TDRI reported that Thailand possesses the scientific and technological personnel capable of tackling various research problems in biotechnology using modern techniques, such as recombinant DNA technology, RFLP mapping, cell fusion, cloning, cell and tissue culture, monoclonal antibody production and other techniques. Yet there has been very little modern biotechnology input into commercializing biotechnology-based products in the main potential industrial sectors, i.e. agriculture, health care, energy and the environment. Examples of the few exceptions are the production of fructose syrup from tapioca using immobilized-cell technology, the waste treatment technology using high-rate-anaerobic processes and a small innovative biotechnology company devoted largely to the production of diagnostic kits.

THE OPPORTUNITIES

The opportunities for utilization in the public and private sector are expected to grow at a fast pace during the next decade. The government through action of the NSTDA action will enhance not only the national R&D but also concentrate on technology transfer and collaboration with the private sector with an aim to commercialization. In terms of funding in the next five years, the national plan is for the government to increase the R&D budget to 0.5% of its annual budget, and the private sector, through various incentives and collaborative schemes, will allocate up to 0.25% of sales in R&D investment. This further investment would definitely lead to increased R&D success and further the development of biotechnology-based industry.

To assist in the development, NCGEB, as the national center, has laid down five major programs, namely:

1. Support of both the private sector and public sector in R&D in genetic engineering and biotechnology.
2. Cooperation with the private sector for the development of biotechnology-based industry.
3. Set up and operation of national laboratories to carry out in-house R&D and to provision of technical services for both private and public sectors.
4. Human resource development in fields related to genetic engineering and biotechnology.
5. Information services and promotional activities in fields related to genetic engineering and biotechnology.

It should be stressed here that in formulating R&D programs, NCGEB uses policy to deliver products to clients, and from them acquires information essential to the formulation of our priorities. From such coordination through numerous round-table discussions, status reports, workshops and seminars, we propose the following biotechnology business opportunities for the next 5-10 years. The opportunities are divided into mission areas, projects and activities using the following criteria:

1. Potential for utilization and commercialization: The activities, especially R&D, should be directed towards well-defined end-users, as much as possible. Consultation with such end-users in industries or the public sector, market feasibility studies, expert judgment, economic analysis of world and local trends in technology needs and other means will be employed to ensure their relevance.
2. Scientific and technological competence and scientific merit: There must already be sufficient scientific and technological infrastructure, especially manpower, to ensure the success of the activity, or the potential to achieve such sufficiency within a reasonable period. Furthermore, the activity should have a sound scientific rationale and have potential to generate new knowledge and know-how, or to strengthen the institutional and manpower base.
3. Comparative advantage in raw materials and other country-specific advantages: Thailand's advantage in having raw materials such as mineral and agricultural resources, and other advantages such as possession of a large pool of skilled

manpower or existing market advantages, should be promoted.

4. Potential in the country's infrastructural development: Emphasis should be made on the technology base of such infrastructural components as telecommunication, energy and rural development. The activities should contribute to the aims of the Seventh Economic and Social Development Plan and other long-term plans for the country's development.

5. Social and environmental impact. Such impact should always be a major factor for consideration, and whenever necessary analysis should be undertaken before launching activities. This includes benefit and risk assessment of direct and indirect impacts on the public (e.g., health, cultural and employment aspects) and on the environment (pollution and conservation aspects).

BIOTECHNOLOGY-BUSINESS OPPORTUNITIES

On the basis of the above criteria and the results of numerous workshops, round-table discussions and status reports, the following six mission areas have been identified:

MISSION AREA 1: BIOTECHNOLOGY FOR PLANT AND PLANT PRODUCT IMPROVEMENT

Sample programs of this mission area are:
a. Varietal improvement
b. Seed/seedling improvement
c. Pest control
d. Post-harvest technology
e. Rice biotechnology

The following targets are expected to be achieved over the next 5-10 years:
a. Papaya varieties resistant to ring spot virus
b. Tomato varieties resistant to leaf-curl virus
c. Cotton varieties resistant to bollworm insects
d. Soybean varieties tolerant to salinity
e. Development of biotechnological techniques that result in longer shelf-life of fruits
f. Basic understanding of control at the molecular level of some important properties of rice, e.g. fragrance.

MISSION AREA 2: BIOTECHNOLOGY FOR ANIMAL AND ANIMAL PRODUCT IMPROVEMENT

Sample projects of this mission area are:
a. Breed improvement
b. Embryo transfer

    c. Animal health
    d. Feed improvement
    e. Marine biotechnology

The following targets are expected to be achieved over the next 5-10 years:

    a. Improved cattle breeds through embryo transfer.
    b. Single-cell protein as protein source in animal feed.
    c. Probiotic production to substitute the use of antibiotic for increase of feed efficiency.
    d. Foot and Mouth disease vaccine production through recombinant DNA technology.

MISSION AREA 3: BIOTECHNOLOGY FOR RURAL DEVELOPMENT AND SMALL FARMERS

Sample projects of this mission area are:

    a. Mushroom cultivation
    b. Algal production
    c. Plant varieties for marginal environments
    d. Agricultural waste utilization

The following targets are expected to be achieved over the next 5-10 years:

    a. Improve mushroom strains and productivity of mushroom production for mass production by small farmers. For example, improve straw mushroom productivity so that average production for hut-cultivation would be higher than 2.5 kg/m^2, and high temperature-tolerant shiitake and champignon mushroom so that mass cultivation by small farmers can be competitive in the world market.
    b. Development of algal production techniques for mass cultivation by small farmers as high-value animal feed and as biofertilizer.
    c. Development of virus-free seed (e.g. potato) and drought-tolerant varieties (e.g. soybean) for the benefit of farmers in marginal land areas.
    d. Improvement of animal nutrition through the upgrading of agricultural and agro-industrial wastes, and through improved fodder crop production.

MISSION AREA 4: BIOTECHNOLOGY FOR NOVEL PRODUCTS/INDUSTRIAL PROCESS IMPROVEMENT

Sample projects of this mission area are:

    a. Development of specialty chemicals, e.g. organic acids, amino acids, food ingredients and other metabolites.
    b. Process improvement and development to improve the

efficiency of current production practices especially for small/medium firms.

   c. Post-harvest technology to improve raw materials' qualities for processing.

The following targets are expected to be achieved over the next 5-10 years:

   a. Development of biosensors for industrial instrumentation, e.g. sensors for ethanol, sucrose, glutamate, BOD and pesticides.

   b. Development of novel fermenters and bioreactors, e.g. high-rate anaerobic digesters, high-productivity/low-energy alcohol fermentation processes.

   c. Strain improvement for cost reduction and quality improvement for the production of chemicals and food, e.g. citric acid, soy sauce, fish sauce, pickled/fermented fruits and vegetables.

   d. Development of processes to produce higher value-added products from starches, e.g. modified starches as food and chemicals.

MISSION AREA 5: BIOTECHNOLOGY FOR SUSTAINABLE DEVELOPMENT

Sample projects of this mission area are:

   a. Development of biofertilizers
   b. Development of biopesticides (micro/viral pesticides)
   c. Mosquito control
   d. Waste utilization and treatment

The following targets are expected to be achieved over the next 5-10 years:

   a. Production of nitrogen-fixing agents e.g. *Rhizobium*

   b. Development and production of biopesticides from toxin-producing microorganisms (bacteria and viruses) for the control of some target insects.

   c. Mass production and formulation of *Bacillus thuringiensis* as a biological-control agent for mosquito control.

   d. Development of a biotechnological process for waste treatment and utilization, e.g. high-rate anaerobic and aerobic treatment.

   e. Waste utilization by biotechnological processes, e.g. production of animal feed from industrial waste.

   f. Development of microorganisms for toxic waste treatment.

MISSION AREA 6: BIOTECHNOLOGY FOR HEALTH IMPROVEMENT

Sample projects of this mission area are:

   a. Health and pharmaceutical products
   b. Development of diagnostic kits for viral diseases
   c. Monoclonal and polyclonal collections

**Table 9.1. Potential Biotechnology Products
for Plant Production and Protection in Thailand**

Type Product	Developer	Stage of Commercialization
Improved varieties		
immunized papaya seedlings	private/public	commercialized
disease-free citrus plants	private/public	developmental phase
disease-free potato plants	public	commercialized (small-scale)
transgenic tomato for virus resistance	public	developmental phase
transgenic papaya for virus resistance	public	developmental phase
orchid plantlets	private	commercialized
eucalyptus seedlings	private	commercialized
Biocontrol agents		
nuclear polyhedrosis virus (NPV)	public	developmental phase
Bacillus thuringiensis	public	developmental phase
Trichoderma spp.	public	developmental phase
Biofertilizers		
Rhizobium spp.	public	commercialized (small-scale)
Mycorrhizal fungi	public	developmental phase
Azolla	public	developmental phase
Blue green algae	private/public	commercialized (small-scale)
Disease diagnosis		
diagnostic kit for cymbidium mosaic virus	public	developmental phase

Potential biotechnology products for plant production and protection in Thailand are presented in Table 9.1.

CURRENT DEVELOPMENTS IN CROP PRODUCTION

GOVERNMENT POLICIES ON CROP PRODUCTION

During 1961-1991 the net agricultural production in Thailand decreased from 40% to 12% and the export value decreased from 90% to 10% in the same period. That means agricultural export commodities such as rice, cassava, rubber, etc., have become less and less important. More and more emphasis has been on industrial and other export commodities. Therefore, strategies on crop production have to be adjusted and placed under four categories as follows:

1. Crops for internal consumption: This category includes cereal crops, oil crops, sugar crops, crops used in the textile industry, multi-purpose trees, herbs and medicinal plants.
2. Crops for export: Some of these may be the same as those in the first group with production surpluses such as rice, cotton, sugarcane, tropical fruits, vegetables, ornamentals, flowers, rubber, cacao, cassava and coffee.
3. Crops for export substitution: There are many crops that Thailand still imports for use in the agro-industry, even though an effort has been made to produce them in the country. Such crops include cotton, soybean, wheat, barley and temperate fruits.
4. Crops which cause potential environmental degradation: In fact all crops seem to do some harm to the ecological environment depending on the types of cropping systems and cultural practices. The slash-and-burn type of cropping in upland areas, for example, has resulted in the loss of fertile land. At present appropriate cropping systems including agro-forestry development have been encouraged.

Thailand's export of major plant biotechnology-based products are presented in Table 9.2 and crop production statistics of the country during 1988-1990 are presented in Table 9.3.

TRADITIONAL AGRICULTURAL RESEARCH

The significance and magnitude of the impact of the so-called Green Revolution was best illustrated by changes in cereal production in India, Pakistan and the Philippines. In both India and Pakistan the rapid increase in wheat yields per hectare has been the major thrust of the Green Revolution. Increases in rice yields also have played a major role in West Pakistan and many other Asian countries including Thailand. The key success of the Green Revolution in Thailand could be attributed to strong and sustained R&D support. The impact, however, may not be as much as that in other countries since Thailand has always been self-sufficient for rice production.

The policy on crop production is to improve productivity and at the same time maintain the ecological environment. Success in crop production in the past has been attributed to strong R&D in variety improvement through conventional or traditional breeding. It is anticipated that biotechnology, especially genetic engineering, will become an important tool in crop production improvement in the future.

Crop variety improvement has been practiced in Thailand for more than 100 years. Rice varieties, for example, were improved through selection by farmers and agronomists during the reigns of King Rama

**Table 9.2. Thailand's export
of major plant biotechnology-based products (1988-1990)**

Product	1988		1989		1990	
	Volume	Value	Volume	Value	Volume	Value
Tapioca products	8,121,532	2,144.0	9,826,153	23,974.0	8,068,885	23,136.0
Rubber	938,701	27,189.0	1,112,488	26,423.0	1,163,802	23,557.0
Rice	5,701,458	34,676.4	6,311,410	45,462.3	4,017,090	27,769.5
Maize	1,212,498	3,828.0	1,181,821	4,093.0	1,235,443	4,144.0
Sugar	1,855,098	9,664.0	2,961,638	19,243.6	2,370,353	17,693.9
Canned pineapple	341,414	4,675.4	345,249	4,399.9	398,337	5,524.0
Mung beans	171,917	1,679.8	128,996	1,274.1	175,431	1,552.6
Fresh orchids	9,545	516.9	10,751	506.8	11,678	552.1
Fresh fruits	54,240	637.0	48,954	573.0	66,134	887.0
Fresh vegetables	42,043	337.0	45,251	484.1	49,276	717.6

Volume : Metric ton; Value : Million Baht

**Table 9.3. Plant area, production and yield
of selected crops in Thailand during 1989-1990**

Crop	1989			1990		
	Area (ha)	Production (1,000 t)	Yield (kg/ha)	Area (ha)	Production (1,000 t)	Yield (kg/ha)
Rice	10,310,240	20,601	2,087.50	9,905,600	17,193	1,956.25
Maize	1,786,400	4,393	2,568.75	1,745,600	3,722	2,406.25
Sugarcane	687,680	33,561	48,893.75	788,640	40,661	51,962.50
Soybean	513,440	672	1,337.50	425,120	530	1,300.00
Cotton	63,840	86	1,362.50	73,760	97	1,362.50
Mungbean	512,800	356	718.75	449,280	303	706.25
Vegetables	1,529,920	20,701	13,918.75	1,491,680	19,705	13,743.75

IV and V (late 1800s-early 1900s). The first rice variety contest was organized in 1907, and the second one in 1909. The country's rice improvement program was initiated soon afterwards. The first elite variety, Pin Kaew, that had undergone selection for 14 years won first prize in the Rice Variety contest in Canada in 1933. Other Thai rice varieties also won 12 out of 20 total prizes in the same contest. These events made Thai rice well-known worldwide, and it continues to maintain its status through today.

Other crop varieties were also improved through crossing and selection for several decades. Thai tropical fruits such as mangoes, mangosteen, pamelo, longan, lichee and durian, and orchids are well-known products of crop improvement efforts. Other economic crops such as rubber, tobacco, cassava, sugarcane, kenaf, cotton, soybean, maize and kapok were introduced into the country for cultivation and at the same time for further improvement. Some of these crops have become major economic crops of the country.

Many government and private agencies have been carrying out crop improvement activities. The Department of Agriculture, Ministry of Agriculture and Cooperatives, for example, has released a total of 149 plant varieties during the past 90 years. Many universities, agricultural colleges and private companies have also released a number of crop varieties. At present Thailand has a well-developed seed industry with many transnational seed companies operating in the country.

SUCCESS STORIES

As a world leader in rice export, the country maintains its rice production capacity through variety improvement and improved cultural practices. More than 30 traditional rice varieties have been released for use in different regions during the past 3-4 decades. These varieties were improved versions of the old varieties with reasonable yield level but good grain and cooking quality. One aromatic rice variety, KDML 105, is very popular among consumers and has a large export demand. Besides the traditional varieties, the Rice Research Institute, Department of Agriculture has collaborated with the International Rice Research Institute (IRRI) since the early 1960s, and many new rice varieties have been released. These new varieties have characteristics of shorter stature, erect leaf, plant pest resistance and high yield potential. However, their grain and cooking quality is not as good as that of the old, traditional varieties.

Maize is considered to be a new crop introduced into Thailand. Maize production has been increasing very rapidly, from less than one million tons a year during the late 1960s to 4.7 million tons in the 1988/89 crop year. The increase in production in the early period was mainly due to the expansion of cultivated areas, but the more recent production increase could be attributed to improvement of varieties and cultural practices. One variety (Guatemala-PB-5) was introduced from Latin America and proved to be adaptable to local conditions. After the outbreak of downy mildew disease to which Guatemala-PB-5 was not resistant, a new downy mildew resistant variety, Suwan 1, replaced it in 1974. The release of Suwan 1 illustrates the successful cooperation among several government and international agricultural agencies. The joint Kasetsart/CIMMYT publication entitled, "Suwan 1: Maize from Thailand to the World" provides a more detailed story.

There are many more crops successfully grown in Thailand as a result of variety improvement coupled with improved cultural practices. Cassava, sugarcane and rubber may be considered among those important export-crop commodities. Ornamental plants such as orchids and cut flowers are also of economic importance. Some temperate crops such as grape and strawberry have adapted quite well to Thailand's conditions after many years of varietal and cultural-practice improvements. These successful crop production examples could be attributed to strong and sustained agricultural research and development.

BIOTECHNOLOGY APPLICATION: THE GENE REVOLUTION

Biotechnology, especially genetic engineering, has developed and advanced rapidly during the past decade. Thailand has joined other countries in searching for ways and means to apply biotechnology for crop production improvement. Since the establishment of the National Center for Genetic Engineering and Biotechnology in 1983, some progress has been made.

TISSUE CULTURE AND MICROPROPAGATION

Tissue culture technology was initiated in the early 1960s and since then the technology has developed well and been used in many areas of plant variety improvement. In Thailand the technology was used in the micropropagation of orchids in 1966. At present more than ten private companies are employing the technology at the commercial level. The export of orchid seedlings and other crops are the result of this development. NCGEB has supported more than 20 R&D projects in this area at various universities since the early 1980s. Many crop species such as rice, maize, soybean, papaya, tomato, potato, strawberry, banana, rubber, rattan, oil palm, teak, and eucalyptus can be multiplied through tissue culture technology.

Tissue culture has been used quite extensively in crop improvement. During the process of propagation many mutant plants have been found and new varieties obtained from this technology. Selection of desirable cells or plants could also be made by subjecting them to various stress conditions. After further improvement selected plants a may be released as a desirable stress-tolerant variety.

The production of disease-free seed potato for commercial sale was also the result of the application of tissue culture for plant disease control. Due to the rapid increase of the area for potato production from 1985 to present, Thailand's importation of seed potato has been increased from 100 tons to 2,000 tons per year. The project was designed to produce disease-free seed potato for commercial use. Although the project has accomplished its scientific objectives as indicated by the well-established procedure for disease-free seed potato production and a diagnostic method for virus detection, commercialization of the production technology has not taken place. There seems to be a lack of an effective policy mechanism to ensure implementation of commercialization.

The private sector would consider this investment a high risk. Moreover, there is no guarantee for companies to be able to recruit sufficient skilled technicians to join the operation. It became clear from this case study that the management of a commercial-scale project requires a highly skilled business-oriented manager for biotechnology-based industries.

GENETIC ENGINEERING: PLANT GENOME MODIFICATION

Genetic engineering or recombinant DNA technology is quite new in Thailand. Conventional plant breeding through the use of crossing and selecting desirable genotypes has resulted in new plant varieties. However, the genetic constitution of these varieties seem to be within a quite definitive set of genes from that species. Modern biotechnology such as genetic engineering allows genetic transformation of genes across species. The process involves culturing and genetic transfer. Culture techniques involve regenerating entire plants from protoplasts, single cells or plant parts. These techniques allow one to grow an enormous number of cells, all potential plants, in a small area, permitting one to screen a large number of cells in a very short time. If one adds a harmful substance (a herbicide, a virus, a disease, salt water, etc.) to the growing media, only those cells resistant to that selective agent will grow, resulting in potentially-resistant plants. The cultured cells also serve as the venue for genetic transformation.

At present the NCGEB Plant Genetic Engineering Unit has the capability to do virus coat-protein transformation in papaya and tomato. Success has been achieved in transforming tomato resistant to tomato yellow leaf curl virus and papaya resistant to papaya ring spot virus. At present genetic transformation techniques are quite well in place and more activities are being carried out.

The designated R&D project entitled "Development and Application of Plant Biotechnological Methods for the Production of Virus Resistant Plants" is now briefly described. According to the proposed workplan, three research areas would be intensively investigated, the first being the genetic constitution of the virus, Tomato Yellow Leaf Curl Virus (TYLCV). The work included virus purification, molecular cloning and mapping of TYLCV coat-protein genes and construction of the chimeric gene with the Ti plasmid. The second area was the establishment of tomato tissue culture emphasizing protoplast isolation, culture and plant regeneration. The third area was the production of transgenic plants. The methods of co-cultivation and leaf disc transformation were developed in order to obtain the transgenic plants.

Two tomato varieties, Seeda and VF134-1-2, were successfully transformed by cotyledon disc transformation. Chimeric gene, pIM603 and pIM606 encoding TYLCV coat-protein genes were used in transformation. The pIM606 construct yielded a higher number of putative transformants than pIM603, and developed faster. The transgenic plants were tested in the greenhouse and proved to be resistant to the tomato yellow leaf curl virus. Similar research work on the production of transgenic papaya resistant to papaya ring spot virus was also successful. The researchers are planning to test these transgenic plant varieties in the field soon with the support of NCGEB.

Molecular marker-aided selection is an important and useful area for crop improvement. The resulting varieties are the same as those obtained from conventional breeding in terms of safety assessment. At present NCGEB in collaboration with IRRI and with financial support from the Rockefeller Foundation has supported rice projects along this line. Efforts are to use molecular markers to aid selection of rice tolerant to flooding and to improve the aromatic quality of rice.

The Rockefeller Foundation has supported the R&D project entitled "QTL Analysis for Submergence Tolerance in Lowland Rice" in cooperation with NCGEB. Since the availability of a complete RFLP linkage map in rice, it is possible to detect genes controlling quantitative traits (QTLs) such as drought resistance, submergence tolerance, fragrance or other quality traits. Data accumulated from sequentially scoring the selective genotype population with informative probes are being used to construct a QTL likelihood map using the MAPMAKER-QTL computer program. The ultimate goal is to utilize the identified RFLP markers flanking the submergence-tolerance QTLs to facilitate improvement of KDML 105 (aromatic rice) and other rice varieties. A similar research project to locate genes conditioning fragrance characteristics has received support from NCGEB.

The DNA fingerprinting technique is a quite useful tool in establishing the identity of plant varieties, lines and hybrids. The NCGEB has funded a project in collaboration with the Rice Research Institute in fingerprinting all lines and varieties of aromatic rice. The activities will serve as a basis for further rice variety improvement and at the same time as the safeguard of Thai rice varieties.

A POTENTIAL MEANS TO PROTECT BIOTECHNOLOGY AND SUSTAINABLE DEVELOPMENT

An important government policy in agricultural development is to reduce the use of chemicals in crop production. The above-mentioned activities of NCGEB comply with this policy as well as in support of the policy of sustainable development. Besides the activities relating to the improvement of plants resistant to various diseases, there are other biotechnology areas that could support the improvement of plant production. The combined resources of conventional technologies and biotechnology coupled with the use of integrated pest management would eliminate over-dependence on agrochemicals, thereby resulting in environmentally friendly development.

Research and development on the production and use of biofertilizers such as *Rhizobium*, mycorrhiza, *Azolla* and bluegreen algae have received much attention. *Rhizobium* and bluegreen algae production have reached the commercial scale. Similarly, R&D on biopesticides such as Bt toxin, NPV virus, neem and others are in progress. Efforts also have been made in identifying Bt genes from local strains that could be used to transform various species of crops such as rice, cotton and vegetables. The resulting varieties will be resistant to major insect pests.

INSTITUTIONAL COLLABORATION

Presently the NCGEB does not have its own laboratories, but facilities will be available in the next three years. In order to strengthen capabilities in specific areas of biotechnology at its collaborative institutes, NCGEB has provided funding and support to specialized laboratories as follows:

1. PLANT GENETIC ENGINEERING UNIT, KASETSART UNIVERSITY, KAMPAENGSAEN CAMPUS

The major task of this institute is to coordinate the various activities of scientists from the affiliated organizations. It also serves as an R&D center for basic and applied research emphasizing economic plants which are resistant to diseases and extreme environmental conditions. Other R&D currently in progress is development of herbicides through biotechnology, strain improvement of papaya resistant to papaya ringspot virus, tomatoes resistant to tomato leaf curl virus and strain improvement and development of viruses for eliminating cotton bollworms.

2. MICROBIAL GENETIC ENGINEERING UNIT, MAHIDOL UNIVERSITY

The Microbial Genetic Engineering Unit is a biotechnology laboratory which emphasizes the study of microorganisms and also disseminates the techniques or technical know-how to all the parties concerned. Recent activities include development of a new strain of bacteria capable of controlling and eliminating disease-carrying insects, particularly a detailed study of a gene responsible for killing mosquito larvae. Studies to insert such a gene into the photosynthetic bacterium *A. quadruplicatum* and to use DNA probes for detecting microorganisms are currently being investigated.

3. MARINE BIOTECHNOLOGY LABORATORY AT SICHANG MARINE SCIENCE RESEARCH AND TRAINING STATION, CHULALONGKORN UNIVERSITY

The main objective of this particular unit is to conduct R&D in aquaculture technology and in marine bioresources for industrial and medical purposes. Current research topics include improvement of oyster species, techniques for abalone cultivation, physiological studies of *Spirulina* spp. and hybridization of king prawn.

4. Biochemical Engineering and Pilot Plant Research and Development Unit, King Mongkut's Institute of Technology, Thonburi

This specialized laboratory aims to be a center for coordinating both scholars and engineers from the network institutions and also serves as a center to strengthen the training programs in biochemical engineering, e.g. design, construction and maintenance of a pilot plant for new biotechnology products or improved production efficiency. Some examples of on-going research are development of biogas production from waste materials of pineapple canneries, production control and analysis of baker's yeast using computers, utilization of CO_2 as a carbon source for *Spirulina* spp. and development of biosensors.

5. Medical Biotechnology Center at Siriraj Hospital, Mahidol University

The objectives of this laboratory are to carry out R&D in the application of biotechnology in health and health related areas. Emphasis will be placed on the development of diagnostic kits for tropical diseases such as dengue and malaria.

6. Microbiological Service Unit, Thailand Institute of Scientific and Technological Research

This unit provides services to public and private researchers for basic research, education and possible commercialization. R&D in microorganism preservation and the utilization of microorganisms in both environmental and industrial concerns are strongly encouraged by this particular organization.

The Plant Genetic Engineering Laboratory has played a key role in developing plant varieties resistant to plant pests through genome manipulation and modification. Other units such as the Microbial Genetic Engineering and Microbiological Service also play supporting roles.

INTERNATIONAL COLLABORATION

NCGEB has collaborative R&D on rice biotechnology with the Rockefeller Foundation (RF) and IRRI. The RF has funded the NCGEB project, "Information Access and Research Management for Rice Biotechnology in Thailand" and two projects granted to Kasetsart University. There are six projects under joint RF/NCGEB funding and eight projects under NCGEB support. NCGEB also funds its own in-house project, "Molecular Studies of Aromatic Rice." There are three sub-projects under this, namely DNA fingerprinting of aromatic Thai rice, development of doubled-haploid rice of KDML 105 (aromatic rice variety), and relationship between 2-acetyl-pyrroline and the fragrance in KDML 105.

In 1993 IRRI initiated the Asian Rice Biotechnology Network (ARBN). There are two components under this collaboration. ARBN-I funded by ADB has three member countries, Philippines, India and

Indonesia. ARBN-II funded by the German Federal Ministry for Economic Cooperation (BMZ) includes two additional countries, Thailand and The People's Republic of China. The objectives of the Network are to promote R&D in rice biotechnology among these five member countries and may later expand to other countries in Asia. Institution strengthening through research support, training, workshops and exchange of visits and materials will be emphasized.

NCGEB has a plan to support collaborative research between IRRI, the Rainfed Lowland Rice Research Consortium, Thai Department of Agriculture, Kasetsart University and the Rockefeller Foundation. The project entitled "Identification, Mapping and Utilization of Rice Blast Resistance Genes in Improved Aromatic Rice Varieties for Northeastern Thailand" will be carried out at the Ubon Rice Research Station. A key objective of this project is to develop commercially-suitable blast-resistant varieties for Indochina and Thailand. This program is poised to participate in the development of required lines, to introduce genes using molecular marker-aided selection, and to evaluate promising lines in a number of different sites. Crosses are being made between aromatic rice cultivars from the region and excellent blast-resistance donors with good grain quality and other desirable traits. These materials are being advanced through anther culture and rapid generation advance to develop a large doubled-haploid population.

PLANT GENETIC RESOURCES

Thailand's economy has rapidly changed from agriculturally-based to more industrial-based. However, agriculture will remain an important sector for decades to come. The capacity to serve as the "Rice Bowl" of Asia has been the result of the government's well-planned agricultural development. Genetic resources management has played an important role in implementing the plan. Genetic resources of the country have been conserved and used very effectively. Many forms of plant germplasm conservation have been practiced in Thailand and include in situ and ex situ conservation, field gene bank, bulk populations, seed gene banks and in vitro gene banks.

Thailand has been actively involved for a long time in plant germplasm collection, preservation and utilization in collaboration with international institutions such as IBPGR, IRRI and CIMMYT (International Maize and Wheat Improvement Center). Collaboration in these activities has proved to be successful, and many crop varieties have been released as the result of these efforts. The released varieties not only benefit Thailand but other countries as well.

INTELLECTUAL PROPERTY RIGHTS

In the final version of the GATT agreement (Uruguay Round), approved on December 15, 1993, Section 3 of Article 27 on "Patentable Subject Matter" states:

"Members may also exclude from patentability:
  a. diagnostic, therapeutic and surgical methods for the treatment of humans and animals;
  b. plants and animals other than microorganisms, and essentially biological processes for the production of plants or animals other than nonbiological and microbiological processes.

However, Members shall provide for the protection of plant varieties either by patents or by an effective *sui generis* system or by any combination thereof. The provisions of this sub-paragraph shall be reviewed four years after the entry into force of the Agreement Establishing the MTO."

The Global Biodiversity Convention which became operational on December 29, 1993 states:

  "reaffirming that States have sovereign rights over their own biological resources," and "reaffirming that States are responsible for conserving their biological diversity and for using their biological resources in a sustainable manner."

Thailand is a signatory to both of these international agreements. Hence, it is essential that legal, administrative and scientific measures are being initiated. The Ministry of Agriculture and Cooperatives and the Ministry of Commerce are taking steps in setting up the "Plant Variety Protection System" that may lead to the Plant Varieties Recognition and Protection Act of the country.

Today intellectual property rights concern almost all nations. Plant Breeders' Rights in particular are of interest. There is a need to use modern technologies such as DNA fingerprinting for the identification of crop lines and varieties in order to protect the country's genetic resources. NCGEB Plant Genetic Engineering Unit has developed techniques that could be used in various crop species, which soon could be of extensive public service.

BIOSAFETY

Biosafety concern is quite new to the country, and it is not widely known even among academia and certain regulatory agencies. The development of biosafety guidelines in Thailand commenced with a commissioned feasibility study by the NCGEB in 1990 after the review of the status report on the prospects of biotechnology in agriculture earlier commissioned in 1986. As a result, the Biosafety Subcommittee was established under the direction of the NCGEB in November 1990 by the Science and Technology Development Executive Board. After the reorganization of NCGEB under the National Science and Technology Development Agency (NSTDA) an ad hoc Biosafety Subcommittee was appointed in April 1992 with the sole responsibility to continue drafting the Biosafety Guidelines. The Draft was completed in June 1992.

Essentially the biosafety guidelines cover guidelines for laboratory practices and the releases of GMOs into the environment. The National Biosafety Committee (NBC) was established in January 1993. An Institutional Biosafety Committee (IBC) at various universities, government departments, research institutes and regulatory agencies as well as private companies also was also strongly recommended.

One of the main tasks confronting the NBC is the launching of a publicity campaign. The final draft of the guidelines has been circulated to various public and private agencies, including the universities and research institutes within the country, for information, comments and criticism. All the comments and criticism received eventually will be taken into consideration and incorporated in the guidelines. The guidelines themselves will be flexible enough to allow modification, revision, updates and amendments whenever deemed essential and necessary in the future.

The guidelines and codes of conduct may be considered soft laws because the adoption of such a guideline or code of conduct is not mandatory, nor is its enforcement observed from the legal point of view. They have no legal binding in the strictest sense. Also, not being statutory or promulgated into laws, questions arise on compliance and legal enforcement. As a result the present biosafety guidelines available in Thailand may be considered voluntary until further legislation is sought and before the guidelines are decreed as a mandatory law. However, several laws exist that could be referred to for the regulation of GMOs in the country. These laws could be utilized in such a way that they could compliment enforcement of the biosafety guidelines whenever necessary.

Implementation and Enforcement of the Guidelines

The initial attempts to implement and enforce biosafety guidelines in Thailand are unique and highly divergent essentially due to the prevalent trend for advanced development of biotechnology in the country. Most research and development projects are largely affiliated with the universities, other academic institutes and government departments dealing with research. A handful of private agencies or companies, both local and subsidiaries of transnationals, also are interested in launching field trials of transgenic plants on various scales. While the biosafety guidelines could be considered soft laws and are nonstatutory at present, the implementation and enforcement at the initial stage will have to rely heavily on existing laws. This kind of approach also has been practiced in industrialized countries, such as various U.S. EPA regulations and other U.S. laws that could govern the releases of GMOs into the environment.

A number of recombinant DNA-related research projects already have been undertaken at several universities in Thailand prior to the conception of biosafety guidelines utilizing locally-generated genetic materials. Although most of the materials investigated to date can be classified as "Generally Recognized as Safe-GRAS," the principal investigators usually adhere to certain safety precautions, and biosafety guidelines developed in other countries relative to the nature of their investigation are observed.

APPLICATIONS FOR FIELD TRIALS OF GMOS IN THAILAND

So far there has been no applications for field trials or releases of GMOs using locally-derived genetic materials. Applications or requests for information for field trials of transgenic materials have been received from Israel, Belgium, U.K. and U.S. In chronological order, they are:

- Development of a vaccine protective against avian colisepticemia in Thailand (Department of Microbiology, Faculty of Life Science, Tel-Aviv University, Tel-Aviv, Israel and Faculty of Veterinary Science, Chulalongkorn University, Bangkok, Thailand). Date of application: June 1991.
- Field evaluation of a new hybrid system in corn (Plant Genetics System N.V., Gent, Belgium). Date of application: August 1992.
- Production of virus-resistant seeds of genetically-engineered cantaloupe and squash (Asgrow Seed Co., The Upjohn Co., Kalamazoo, Michigan, U.S. and the Upjohn Co., Pang khon, Sakon Nakhon, Thailand). Date of application: September 1992.
- Seed production of Calgene Flavr Savr tomato was proposed by Thai-Pan Company (local company) in July 1993. It was the only case that received permission from the NBC to carry out the field trial under containment (net house) in late 1993-early 1994. The trial served as a case study for biosafety monitoring practices, and all activities went well under NBC supervision.

It is anticipated that a greater number of applications for GMO field testing will be received in 1994.

REFERENCES

1. Agricultural Statistics of Thailand Crop Year 1990/91. Bankok: Office of Agricultural Economics, Ministry of Agriculture and Cooperatives, 1991.
2. Bhumiratana S. Biotechnology-Business Opportunities in Thailand. Proceedings of the 1992 NSTDA Annual Conference, Cholburi, 4-6 September, 1992.
3. Chulavatanatol M. Status of R&D in Thai Biotechnology. Proceedings of STDB Seminar on Business Opportunity in Biotechnology, Pattaya, Cholburi, Thailand, 3-5 October, 1990.
4. Napompeth B. Biosafety Regulations in Thailand. Proceedings of the ISAAA Biosafety Workshop, Cisarua, Bogor, Indonesia. 19-23 April, 1993. Ithaca: ISAAA, 1993.

5. Senanarong A. Biotechnology for the improvement of plant varieties and products. In: TDRI Reports "R&D Biotechnology for the Improvement of Socio-economic conditions of Thai People". Bankok: TDRI, 1994.

6. Sriwatanapongse S, Jinahyon S, Vasal SK. Suwan 1: Maize from Thailand to the World. Mexico, DF: CIMMYT, 1993.

7. Sriwatanapongse S. Biodiversity Conservation and Utilization: Thailand's Experience. In: "Widening Perspectives on Biodiversity". Geneva, Switzerland: International Academy of the Environment, 1994.

CHAPTER 10

PLANT BIOTECHNOLOGY IN CHINA

Hongya Gu and Zhangliang Chen

INTRODUCTION

China is the largest agricultural country in the world with a population of about 1.2 billion, of which two thirds are farmers. In a country with only 7% of the world's cultivable land, feeding 20% of the world's population is a gigantic "project." The application of traditional biotechnology in China can be traced more than a thousand years, back to when our ancestors used it to produce soy sauce, vinegar, wine and other fermented food. Chinese people also have accumulated a very rich experience in traditional plant breeding. The success of semi-dwarf and hybrid rice varieties, bred during the 1950s and 1970s increased rice yield substantially, and the area of cultivation for these rice cultivars reached 16 million hectares.

Chinese scientists paid much attention to the upsurge of plant biotechnology in the 1960s and 1970s, and many of them took a very active part in research work on plant biotechnology, especially on tissue culture, cell culture and protoplast culture. Their achievements not only made China a major country for tissue culture, but also contributed a great deal to the development of plant biotechnology in the world. Many of them were invited to laboratories of developed countries to disseminate the techniques of plant tissue culture. From the mid 1970s on, molecular biology and modern biotechnology have been developing rapidly in the world, and the Chinese government attaches great importance to it.

In 1986, the government launched National High Tech Planning or 863 Planning (named for March of 1986 when the Planning was initiated) with seven high technology fields selected as national priorities.

Plant Biotechnology Transfer to Developing Countries,
edited by D.W. Altman and K.N. Watanabe. © 1995 R.G. Landes Company.

Biotechnology ranks as the top one. Among three areas in biotechnology, plant biotechnology has the first priority. Therefore, over 100 laboratories are strongly supported by grants from 863 Planning with research projects covering most aspects of plant biotechnology: tissue, cell and protoplast culture; hybrid breeding, embryo rescue and RFLP mapping for breeding; genetic engineering to obtain transgenic plants with resistance to viral, fungal and bacterial pathogens, pests and stress, and with high yield and better quality; and the Rice Genome Project. Biotechnology for herbicide resistance was not considered a priority because of sufficient labor available in the countryside. Besides 863 Planning, the governmental 5-Year Planning has provided strong financial support for plant biotechnology. Some international organizations such as the World Bank and the Rockefeller Foundation and some foreign governments have helped Chinese scientists in developing plant biotechnology. All of these have enabled China to take a leading role in this field among Asian and developing countries.[1,2]

PLANT BREEDING

China has a long history of rice cultivation. The three-line system of breeding hybrid rice[3] is a major contribution to rice breeding in the world. The discovery of a photoperiod-sensitive genetic male-sterile rice cultivar in Hubei Province made it possible to bring three-line hybridization down to two-line. Some hybrid rice plants bred from the two-line system have been cultivated in many areas in the country. However, this system is limited by instability of male sterility. Chinese scientists are also trying another way to reduce the number of breeding lines in rice hybridization and fix hybrid vigor: to obtain apomictic rice.

Chinese scientists have also successfully bred some rice cultivars from anther or pollen cultured in vitro, such as the Zhong-hua series. This series is now well-known for its high yield, better quality and resistance to certain pathogens and is now widely planted in China. In addition to rice, about 20 different haploid plant varieties have been obtained from anther culture.

Plant regeneration from protoplasts is a very important technique in plant biotechnology. Researchers in China have regenerated plants from protoplasts of rice, wheat, millet, soybean, *Brassica*, *Medicago* and other species. At present, they are at the stage of introducing target genes into rice plants.

ANTI-VIRUS RESEARCH WITH PLANT BIOTECHNOLOGY

When Beachy's group[4] found that tobacco plants transformed with the coat-protein gene of Tobacco Mosaic Virus (TMV) could resist infection, Chinese biologists immediately followed up this discovery, started to isolate coat-protein genes from Chinese strains of pathogenic viruses and tried to transfer them into plants. This has become the most active subject for plant biotechnological research in China.[1] For example, research workers have cloned and characterized coat proteins of TMV,[5] Cucumber Mosaic Virus (CMV),[6] Potato Virus X[7] and Y,[8] Soybean Mosaic Virus,[9] Rice Dwarf Virus,[10] Papaya Ring Spot Virus and so on. Some of the genes have been integrated into the genome of certain crops such as tobacco, tomato, potato, green pepper, chrysanthemum and rice, and field tests have been conducted in various regions in China. PK873, the tobacco plants transformed with the TMV coat-protein gene, were field tested on about 800 hectares in 12 provinces of China during the past few years.[11] Transgenic tomato with a coat-protein gene from CMV is also at the stage of field testing.[12] Researchers in the Institute of Microbiology, Chinese Academy of Sciences have transferred TMV or/and CMV coat-protein genes into two popular tobacco cultivars, NC89 and SD-8703, and the results from their field test showed that the transgenic tobacco plants could resist TMV and/or CMV infection.

Besides coat-protein genes, Chinese scientists are trying to use other genetic components from viruses to transform plants, such as satellite RNA, or use genes coding for replicase or movement proteins to challenge the transgenic plants with the virus to test their resistance. Another approach to control viruses is to transform plants with resistant genes isolated from plants which show resistance to viruses under natural conditions. Trichosanthin (TCS) is a good example. TCS is a protein found in the Chinese medicinal herb *Trichosanthes kirilowii* with the activity of Ribosome Inactivating Protein (RIP). It has been shown that TCS can inhibit the growth of some human viruses including HIV. Recently, researchers at the National Laboratory of Protein Engineering and Plant Genetic Engineering of Peking University cloned and characterized the gene encoding TCS and have expressed it in *E. coli*, yeast and tobacco.[13] When they challenged the transgenic tobacco plants with viruses, they found that the plants could resist viral infection to a certain extent.[14]

ANTI-BACTERIA AND -FUNGI RESEARCH

Bacteria and fungi are major plant pathogens. Rice bacterial blight alone could cause a 10% reduction in rice yield. The Chinese have a long history of using traditional methods to breed crops with resistance to these pathogens. Following advances in the development of plant biotechnology, Chinese scientists are introducing RFLP-mapping techniques into plant breeding programs for locating pathogen-resistance traits and possibly for isolating the resistance genes. At the same time, researchers have screened over 1,000 antagonistic bacteria and many plants in China for promising genes, rich natural resources, or their specific products that have the function of inhibiting the growth of pathogenic bacteria or fungi. Now, they have purified and sequenced several new proteins from bacteria[15-18] and rice[19] and found that these proteins can strongly inhibit the growth of certain bacteria and fungi. Genes encoding some of these proteins also have been cloned and sequenced. The final goal is to transfer these genes into plants to make them resistant to pathogens.

INSECT-RESISTANCE RESEARCH

Chinese scientists have cloned genes encoding an endotoxin from *Bacillus thuringiensis* and a trypsin inhibitor. Both of these genes have been transferred into some plants, and it was found that the transgenic plants can resist insects. But, up to now, Chinese scientists have not found an efficient way to control cotton bollworm and rice brown hopper, which have gained resistance to various chemical insecticides. They are trying very hard to cooperate with foreign colleagues to find efficient methods to control these pests. Researchers at Peking University and Beijing Agricultural University have transferred both the endotoxin gene of *Bacillus* and the coat-protein gene of TMV into tobacco plants and found that transgenic plants could confer resistance to both insects and TMV infection.[20]

NITROGEN-FIXATION RESEARCH

Nitrogen is an essential element for plants, but uptake is principally in the form of nitrate. Some bacteria in soil can fix atmospheric nitrogen and thereby benefit plants growing with them. *Rhizobium leguminosarum*, the bacterium symbiotically associated with the root of legumes to cause nodules, is one of them.

Currently, there are two approaches in China to increase nitrogen in the form available for plants in soil. One approach is to study the mechanism of nitrogen-fixation at the molecular level, the other is to make recombinant *Rhizobium* strains highly efficient in nitrogen fixation. These strains have been spread on more than a million hectares of rice and soybean fields and have increased the yield from 5% to 10%, depending on the condition of the soil. Final evaluation of this application of biotechnology is still pending because the biosafety issue has been raised.

CROP-QUALITY IMPROVEMENT
AND STRESS-RESISTANCE RESEARCH

Traditional plant breeding for obtaining crops with better quality and high yield has been practiced by the Chinese for a long time. It is now supplemented by modern techniques. Some genes coding for plant storage proteins have been cloned, characterized and transferred into crops. Other projects are involved in increasing the starch and oil content of crops and improving their quality. Stress-resistance research is focused on saline and drought resistance. Large areas of farmland are composed of saline soil in China, and several groups of scientists are trying to map and clone the saline-resistance genes.

ENVIRONMENTAL PROTECTION WITH PLANT BIOTECHNOLOGY

Farmland polluted by heavy-metal elements is a severe problem in China. A research group in Peking University recently cloned a human metallothionein (MT) gene and transferred it into tobacco plants with expression driven by a plant promoter. In the experiments, cadmium (Cd) was fed to normal and transgenic plants. The former were killed when Cd concentrations reached 10 µM while the latter group could still grow at concentrations of 250 µM. More importantly, the transgenic tobacco plants can absorb Cd from soil because of the presence of metallothionein in their cells.[21] Researchers are now transferring this gene into weeds and in hopes of using the transgenic weed as a heavy metal cleaner for farmland. They are also using a root-specific promoter to drive this gene in order to localize heavy-metal elements in the root tissue of transgenic crops so that farmers can harvest seeds without heavy metal contamination.

RICE GENOME PROJECT

In August 1992, the Chinese government announced the initiation of a national project—the Rice Genome Project. This is a 15-year project headed by Professors Guofen Hong and Zhangliang Chen. Currently, six institutions in China are participating in this project, including a center and five satellite laboratories. This project involves genetic and physical mapping of the rice genome, making various cDNA and genomic libraries including YAC or BAC libraries, and sequencing the complete rice genomic DNA of about 4.6×10^8 base pairs. Thus, strong collaboration and exchanges with other countries are necessary.

BIOSAFETY CONTROL ON BIOTECHNOLOGY IN CHINA

With strong financial support from the Chinese government as well as support from foreign agencies such as the Rockefeller Foundation, UNESCO, UNIDO, the European Community and others, many Chinese institutions and scientists are involved in biotechnology research and development. Through their efforts, the following have occurred: the first genetically-engineered pharmaceutical product in China, interferon, has been put into production on a large scale; the recombinant vaccine for Hepatitis B Virus has completed second-phase clinical trials; the interleukin-2 gene has been cloned and expressed in *E. coli*; and some transgenic, anti-virus plants or genetically-altered microorganisms have been tested in the field with some commercial products made from them. In addition, three kinds of engineered vaccines against acute diarrhea in baby pigs have been successfully developed and tested. As more and more engineered pharmaceutical-biotech and agricultural-biotech products are and/or will be commercialized for the Chinese market, the government is paying much attention to the biosafety issue as well.

In 1990, the Chinese government decided to set up a legal system for the application of genetic-engineering techniques and its biosafety control. The State Commission of Science and Technology, together with the Ministry of Public Health, Ministry of Agriculture and Chinese Academy of Sciences, established a Special Committee to formulate the Regulation. Madam Zhu Li-an, Deputy Managing Director of the State Commission of Science and Technology, was appointed the Chairwoman of the Special Committee. Later on, a Consultation Committee, an Integrated Drafting Group and an administrative unit were set up for deliberating and drafting the Regulation.

The Integrated Drafting Group has consulted and referred to more than 40 national and international regulations related to the biosafety control on genetic-engineering manipulations, and called for expert meetings many times. It has also sent questionnaire forms to more than 300 experts across China, and interviewed scientists in research institutions of the Ministry of Public Health, State Bureau of Medicine and Pharmacology, Ministry of Agriculture, Ministry of Light Industry, Bureau of Environment Protection and others to ask for their opinions, comments and suggestions. Based on the results of these surveys, the preliminary draft of the Regulation was outlined in July 1990. After being studied and modified many times, in March 1991 the final draft was sent to the Special Committee, Consultation Committee and to relevant governmental or scientific authorities, such as the Ministry of Public Health, Ministry of Agriculture, State Bureau of Medicine and Pharmacology, Ministry of Light Industry, Bureau of Environment Protection, State Commission of Education and Chinese Academy of Sciences for more suggestions and comments. Based on the feedback from those authorities and scientists, the Regulation (draft) has been modified and revised again several times. It was approved by the Special Committee on 11 November 1992. The approved Regulation was sent by the State Commission of Science and Technology on 30 January 1993 to the Chinese State Council for final ratification. It was activated by the order issued by Mr. Song Jian, the Chairman of State Science and Technology Commission on 24 December 1993. It is promulgated nation-wide and now serves as a general guideline for biosafety control on genetic engineering work in China.

CONCLUSIONS

In summary, plant biotechnology is extremely important in China, a very big agricultural country. With support from the Chinese government and some international organizations, and the efforts of Chinese scientists, China has made certain achievements in this field. There are several areas or aspects that need more improvement, however. For example, basic studies on molecular biology are not sufficient, and studies on gene regulation are relatively weak compared with those in developed countries. Also, there has not been enough funding and manpower for research in or development of molecular biotechnology. Like many developing countries, China is facing a severe problem of brain drain; many qualified scientists and well-trained students go to developed countries due to insufficient research funds and/or low salaries in their homeland. Another issue is that China should pay more attention to biosafety issues for the release of genetically-engineered organisms. It should be pointed out that the Chinese government has formulated a regulation for controlling and managing the release of transgenic organisms in China and is seeking more cooperation with foreign organizations, research institutions and scientists in this area.

REFERENCES

1. Coghlan A. China's new Cultural Revolution. New Scientist 1993; 137:4.
2. Chen ZL. Application of Bioengineering to Agriculture. High Technology Newsletter 1992; 2: 1.
3. Yuan NP, Chen HX, eds. Breeding and Cultivation of Hybrid Rice, Hunan: Hunan Science and Technology Press, 1988:65-155.
4. Powell AP, Nelson RS, Barun De et al. Delay of disease development in transgenic plants that express the tobacco mosaic virus coat protein gene. Science 1986; 232:738-740.
5. Tian YC, Qin XF, Wang GL et al. Virus resistance of transgenic tobacco plants expressing Tobacco Mosaic Virus coat protein gene. Scientica Sinica Series B 1991; 34(11):1329-1340.
6. Hu TH, Wu L, Liu W et al. cDNA cloning of gene encoding coat protein of CMV infecting potato plants in China. Chinese Science Bulletin 1989; 21:1652.
7. Wang CX, Gao Q, Pan NS et al. The cDNA cloning and nucleotide sequence of potato virus X coat protein gene. Acta Botanica Sinica 1991; 33:363-369.
8. Chu RY, GUO T, Pan NS et al. Cloning and sequencing of the gene encoding coat protein of potato virus Y. Acta Botanica Sinica 1992; 34:191-196.
9. Chu RY, Leng XH, Bao YM et al. Amplification of soybean mosaic virus coat protein gene by polymerase chain reaction and its sequence analysis. Acta Botanica Sinica 1992; 34:523-528.
10. Chu RY, Zhang X, Pan NS et al. The cDNA cloning and nucleotide sequence of the gene encoding nonstructure protein of rice dwarf virus genome segment 10. Acta Botanica Sinica 1993; 35:115-120.
11. Chen ZL, Sun BJ, Pan NS et al. PK-873, an excellent virus-resistant transgenic fragnant tobacco. In: Chen ZL, ed. Studies on Plant Genetic Engineering, Beijing: Peking University Press, 1993:219-226.
12. Yang, RC. Collected papers on Celebration of Sixty's Anniversary of Jiangsu Academy of Agriculture. Jiangsu: Jiangsu Academy of Agriculture, 1992:19-23.
13. Bao YM, Chu RY, Han JH et al. Cloning and sequencing of trichosanthin gene and its expression in *Escherichia coli* and tobacco plant. Scientica Sinica Series B 1993; 36(6):669-676.

14. Zhou P, Zhu YX, Zheng WH et al. Prevention of disease development in transgenic plant expressing the trichosanthin gene. Asia-Pacific Journal of Molecular Biology and Biotechnology 1993; 1(1):57-63.
15. Liu JY, Li N, Pan NS et al. Purification and partial characterization of an antibacterial protein LCIII. Chinese Journal of Biotechnology 1992; 8:266-271.
16. Liu JY, Liu W, Pan NS et al. The characterization of antagonistic bacterium AO14 and its antibacterial protein. Acta Botanica Sinica 1991; 33:157-161.
17. Wang YP, Liu YQ, Pan NS et al. Studies on the biological control of plant diseases with *Bacillus subtilis* strain TG26. Chinese Journal of Biological Control 1993; 9: 63-69.
18. Wang YP. Agricultural biotechnology. In: You CB, Chen ZL, eds. Biotechnology. Beijing: China Science and Technology Press, 1993:789-793.
19. You CB, Chen ZL, Ding Y, eds. Biotechnology in Agriculture. Netherlands: Kluwer Academic Publishers, 1993:11-17.
20. Liang XY. Studies on Anti-insect Virus with Plant Genetic Engineering, Ph.D. Dissertation, Beijing Agriculture University, 1993.
21. Pan AH, Tie F, Yang MZ et al. Construction of multiple copy of 5'-domain gene fragment of human liver metallothionein IA in tandem arrays and its expression in transgenic tobacco plants. Protein Engineering 1993; 6(7):755-762.

ABOUT THE AUTHORS

Zhangliang Chen received his Ph.D. degree from Washington University in St. Louis, MO, U.S. in 1987 and is now the Dean and Professor in the College of Life Sciences at Peking University, Beijing. He is one of nine members in the China National High Technology Planning, Biotechnology Consultant Committee and Advisor in the Advisory Committee for Biotechnology of China's Agriculture Ministry. He is the 1991 winner of UNESCO's Javed Husain Prize for Young Scientist.

Hongya Gu received her Ph.D. degree from Washington University in St. Louis, MO, U.S. in 1987 and is now a Professor in the College of Life Sciences at Peking University, Beijing, China.

AN EXAMPLE OF TRANSFER OF PROPRIETARY TECHNOLOGY FROM THE PRIVATE SECTOR TO A DEVELOPING COUNTRY

Rafael Rivera-Bustamante

POTATO CULTURE IN MEXICO

Potato is one of the most important crops in Mexico and ranked fifth on the priority list (after maize, beans, wheat and cotton) of the National Institute for Research in Forestry, Agriculture and Livestock (INIFAP), the research agency of the Ministry of Agriculture and Water Resources. In 1992, potato occupied approximately 72,000 ha with an average production of 16.8 tons per hectare. Total annual production was in the range of 1.21 million tons and valued at $240 million. Approximately 80% of the production is used for consumption as table potatoes (fresh market), 7% for industry (potato chips, flour, starch and alcohol production) and 13% for seed.

It is noteworthy that almost 57% of the production is under irrigation where yields average 20.5 tons per ha; the remaining 43% is rainfed with average yields of 11.7 tons per ha. The major production areas are in the states of Sinaloa (11,800 ha, 100% irrigation), Puebla (11,400 ha, 20% irrigation) and Mexico (8,500 ha, 39% irrigation). Other important areas include the states of Chihuahua, Guanajuato, Coahuila and Nuevo León.[1]

Alpha, an old Dutch variety, occupies approximately 60% or 45,000 ha of the total 80,000 ha in the country. The second most important variety is López (15,000 ha), and the remaining acreage is occupied by various Mexican varieties and others such as Atlantic (chipper), White Rose and Bintje.

Alpha is a variety that is very well adapted to Mexican conditions and is grown in all states at altitudes from 20 to 3,200 meters above sea level. This variety is very popular because of its versatile qualities for cooking, as a chipper or for industrial use to produce flour, starch and alcohol. Despite its extreme susceptibility to late blight (*Phytophthora infestans*) and virus diseases, Alpha has retained its position as the major variety for many years, and there is no indication that it will be replaced in the short term.

Two types of potato farmers can be easily distinguished in Mexico. Industrialized farmers with modern, irrigated and well-organized farms utilize certified seed produced by themselves or through their own farmer organizations. Phytopathological problems are usually under control with heavy applications of chemical products and integrated pest management. On the other end, there are small farmers that rarely use certified seed, produce potatoes under nonirrigated conditions and can not afford a high number of chemical applications. Their potato seed is usually the remains of the previous year's harvest or the ones that could not be sold in the market because of small size or low quality.

Potato is not a typical "commercial" crop in Mexico; there are no exports, and imports are minimal. Average price for ware potato in the market place ranges from 10 cents to $1.00 per kilo, with seed potatoes valued at 25 cents to 65 cents per kilo. These variations are often discouraging for potato-seed producers. They can not add a premium to their certified seed because high seed prices in comparison with table potatoes will drive farmers to buy potatoes at the fresh market and to use them as seed. This is especially evident during a year or season with high production and a low market price.

GENETIC ENGINEERING POTATOES
FOR VIRUS RESISTANCE

Improvement of potato by breeding and selection is laborious and inefficient because it is a tetraploid species. In addition, the new varieties obtained, although incorporating new important characteristics, usually compete at a disadvantage with the well-established varieties that farmers have been growing for decades (e.g. Alpha in Mexico and Russet Burbank in U.S.). Genetic engineering has been shown to be an excellent tool to improve existing varieties by selectively adding specific traits such as virus resistance. Genetically-engineered resistance to virus disease was one the first practical demonstrations of this powerful technology.[2] Several new strategies to obtain virus-resistant plants are now available and have been applied to almost all types of plant viruses, including the most important potato viruses.[3]

In 1990, Monsanto researchers reported transgenic Russet Burbank potatoes resistant to mixed virus infections.[4] This was the first attempt to obtain coat protein-mediated protection against two different viruses (PVX and PVY) with a single transformation event. The vector used, pMON9898, contained two chimeric genes, each containing a coat-protein gene (PVX or PVY) under the control of the 35S promoter and a termination signal. The selectable-marker gene, NPTII that confers resistance to kanamycin, also was included. With this vector, Monsanto researchers were able to obtain several potato lines that expressed both coat-protein genes and were resistant to infection by PVX and PVY after mechanical inoculation. One line (#303) was also resistant to PVY when inoculated with viruliferous aphids. Four selected lines were field tested for virus resistance, plant growth and tuber yield.[5] Line 303 was highly resistant to both viruses, and after the double inoculation tuber yield was unaffected whereas in nontransgenic controls it decreased. The tests also showed that uninoculated transgenic clones had tuber yields as high as control Russet Burbank. Since then, several field tests at different locations have confirmed that resistance to PVX and PVY is effective in the field and can prevent yield losses due to dual infections by these viruses.

MEXICO-MONSANTO COLLABORATION PROJECT

A collaborative program, between Monsanto and Mexico to introduce coat-protein genes into Mexican potato varieties and to confer genetically-engineered resistance to potato viruses PVX and PVY, was initiated in January 1991. This program is the first project sponsored by ISAAA to transfer technology between a private company and a developing country. ISAAA identified the Mexican needs in potato culture, contacted Monsanto as owner of the technology and found financing for the project from The Rockefeller Foundation.

The project is expected to have impact in five areas:

A. *Productivity.* An increase in potato production is expected by controlling virus diseases. Because the technology is incorporated into the "seed," it has the capability of reaching all types of farmers and of having an impact at all levels.

B. *Environment.* A potential environmental impact is anticipated by decreasing the need for heavy insecticide sprays that are common in some areas of potato cultivation. This also will reduce production costs and could be an important contribution to sustainable agriculture.

C. *Technology Transfer.* The project can be considered as a pilot program and a unique model for exploring the mechanisms that will facilitate the sharing of proprietary-biotechnology applications by industry with developing countries.

D. *Biosafety Procedures.* The project also represents a vehicle that will facilitate the establishment of regulatory procedures and the development of biosafety guidelines for testing and introducing recombinant DNA technology in Mexico.

E. *Personnel Training.* The project provides an opportunity for Mexican scientists to participate in hands-on training at Monsanto and to return to their own institution (CINVESTAV) to establish the learned methodologies.

Table 11.1 summarizes the four phases of the project. The objectives of each phase are described in the following section in combination with the results obtained up to June 1994. Because training of personnel is a fundamental part of any project involving transfer of technologies, the following section will emphasize that aspect.

DESCRIPTION OF THE PROGRAM

Phase One:

This phase was programmed to be carried out at Monsanto facilities in St. Louis, Missouri, U.S. with the following objectives:

1. Training of CINVESTAV personnel in the protocols for the transformation and regeneration of the potato variety Russet Burbank.
2. Adaptation of Russet Burbank transformation protocols to the Alpha variety.

Phase Two:

The objectives for this phase, also in St. Louis, were:

1. Training in gene expression and viral-protection assays for analysis and screening of transgenic plants.
2. Construction of an improved vector (pMON18770) containing PVX and PVY coat proteins.

Table 11.1. Calendar of activities for Mexico-Monsanto potato project

Activity	1991	1992	1993	1994
Phase one				
Training in potato transformation technology (Monsanto)	XXXX			
Establishment of technology at Irapuato		XXXX		
Phase two				
Molecular virology		XXXX		
Phase three				
Transformation at Irapuato (Alpha variety)		XX	XXXX	XXXX
Laboratory analysis and testing of transformants		X	XXXX	XXXX
Phase four				
Field test (small-scale)			XXXX	XXXX
Field test (large-scale)				XXXX

Phase Three:

The objective in this phase is to establish at CINVESTAV (Irapuato, Guanajuato, Mexico) the methodologies for tissue culture and transformation of potato as well as the protocols for analysis and evaluation of transgenic materials.

Phase Four:

The objectives of this phase are:

1. Exposure of CINVESTAV personnel to field-evaluation procedures for transgenic plants. This includes participation in Monsanto's field trials with PVX/PVY and PLRV resistant potatoes.
2. Exposure to the regulatory issues in conducting field tests.
3. Field testing of transgenic Alpha-potatoes.

RESULTS AND CURRENT STATUS OF THE PROJECT

For Phase I, a scientist from CINVESTAV (Rosa Ma. Rangel-Cano) joined Monsanto's Crop Transformation Group at St. Louis in April 1991.

During the first three months Ms. Rangel-Cano worked with the potato variety Russet Burbank to familiarize herself with the system. The methodologies for transformation and regeneration of this variety are well-established at Monsanto, and results already had been published, including selection of several potato lines with resistance to PVX and PVY.[4,5]

After training with Russet Burbank, Ms. Rangel-Cano dedicated her work to the Alpha variety. Because the methodologies were developed specifically for Russet Burbank, a few parameters had to be verified with Alpha to assure an equivalent transformation efficiency (e.g., kanamycin resistance, inoculation and cocultivation times, etc.). To verify each parameter experiments with an average of 1,000 explants were carried out. Because efficient conditions for Alpha were similar to those already established for Russet Burbank, it was decided to use the same protocol. The results for Alpha transformation are summarized in Table 11.2.

Results for the transformation of Alpha variety could be summarized as follows:

1. An average of 50% of the explants produced shoots.
2. From the total number of shoots obtained, an average of 63% were positive for the NPTII recallusing test.
3. From the plants that developed callus in kanamycin media, only 50% were confirmed by NPTII ELISA (pMON9898) and 45% by GUS assay (pMON15722-9).
4. The final efficiency of transformation is in the range of 16% (average of all transformation experiments with Alpha).

Interestingly, the efficiency obtained with Alpha was 4-5 times higher than that obtained by Monsanto with Russet Burbank using the same

Table 11.2. Efficiency of transformation* for potato variety Alpha with different vectors

Kanamycin concentration	Vector used (pMON#)	Number of Explants	% Explants w/shoots*	% Positive recallusing**	% Positive (GUS or NPTII)***
200	15722-9	420	55.3	62.3	57.0
200	9898	1,000	42.8	62.3	49.4
100	15722-9	491	67.2	65.7	32.4
100	9898	1,106	49.2	65.7	50.2

* Potato explants were transformed with either pMON15722-9 (GUS-containing vector) or pMON9898 (vector with PVX and PVY coat-protein genes). After co-cultivation with *Agrobacterium tumefaciens,* the explants were cultured with two different antibiotic concentrations (100 or 200 mg/ml kanamycin).
** Recallusing: callus formation in media with kanamycin to verify presence of NPTII gene
GUS: b-Glucuronidase assay
*** NPTII; ELISA assay with antibodies against neomycin phosphotransferase protein

vector (pMON9898). From all plants that were positive for ELISA-NPTII (using pMON9898), 108 independent lines also were tested for expression of PVX coat protein by ELISA. The 50 highest-expressing lines for PVX coat protein were further selected and propagated for field testing in Prosser, Washington, U.S.

For Phase II, a second Mexican scientist (J. Trinidad Ascencio-Ibañez) joined Monsanto's virology group in November 1991. His training included transient-expression assays (protoplast), Northern and Western analyses, viral-protection assays on transgenic plants, virus purification, virus inoculation and evaluation of viral-disease development.

The second objective in this phase was the construction of an improved version of the PVX and PVY coat protein-gene vector. Although the pMON9898 vector had been used successfully by Monsanto to obtain transgenic potato plants resistant to PVX and PVY, it is basically an experimental vector that is not suitable for obtaining products with commercial value. A new vector (pMON18770) shown in Figure 11.1 was constructed using the same chimeric genes present in pMON9898. Plasmid pMON18770 is 11,796 base pairs long and contains the coat-protein genes of PVX and PVY along with the enhanced version of the 35S promoter from cauliflower mosaic virus (35S CaMV). It also contains the NPTII gene for kanamycin resistance. All of these segments are between the right and the left borders of the t-DNA, thus the material that will be transferred to the plant is perfectly delimited. As a selectable marker in bacteria, pMON18770 also contains a gene that confers resistance to spectinomycin and streptomycin. This vector was used to obtain potatoes to be released to the farmers. The transformation of Alpha potatoes with this vector will only be done at CINVESTAV.

Phases Three and Four are currently being carried out. The protocols for tissue culture and transformation of Alpha potatoes already have been established. New transgenic lines have been obtained with pMON18770, and a field test is programmed for 1994-95 to evaluate most of these lines. The test will be performed in collaboration with INIFAP personnel. Conditions for tissue culture and transformation of other Mexican potato varieties are being evaluated. These varieties (Mexiquense, Norteña and Rosita) are candidates to be included in the program (discussed in the next section).

Because Phase I of this project was successfully implemented and transgenic-Alpha potato were becoming available, a field test was scheduled. The test was conducted during the summer of 1992 in one of the locations where Monsanto carries out field tests with transgenic potatoes (Prosser, Washington, U.S.).

Thirty lines were tested in a design requiring 40 (4 x 10) replicates of each line. Each plant was mechanically inoculated with PVX, whereas for PVY, spreader nontransgenic plants inoculated with PVY were interspersed with the transgenic plants. Between 20-25 aphids were deposited on each spreader plant to facilitate PVY dissemination. The levels of PVX and PVY infection were evaluated at the end of the season by analyzing each plant with an ELISA assay for PVX and PVY. The percentages of infection were calculated, and the best lines are shown in Table 11.3. This test was done in collaboration with a virology group at Prosser (headed by Dr. P. Thomas, Washington State University). These experiments exposed CINVESTAV personnel to field-testing design, the handling of transgenic potatoes including adaptation after hardening off from in vitro conditions, open release to the environment and evaluation of resistance levels.

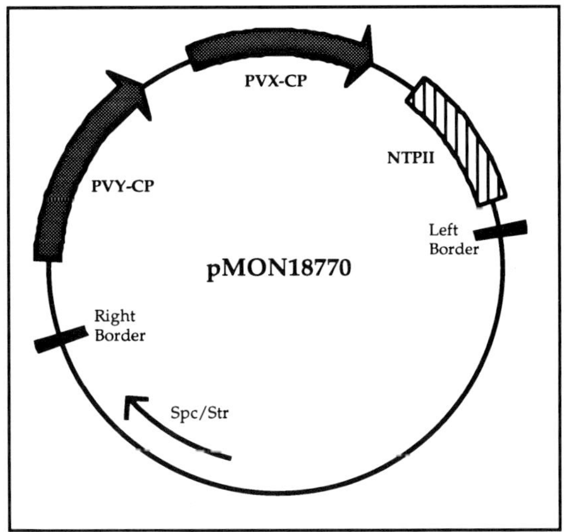

Fig. 11.1. Simplified diagram of plasmid pMON18770. An improved vector was constructed as part of the potato project. In this vector, the t-DNA (DNA that is actually transferred to the plant genome) is restricted only to cpasid-protein genes (PVY-CP and PVX-CP) needed to confer virus resistance and the kanamycin-resistance gene (NPTII) required to select the transformed cells. The t-DNA is delimited by the Right and Left Border. Outside the t-DNA remains a fragment (Spc/Str) required to facilitate the selection of bacterial cells containing this plasmid.

Table 11.3. Field test results of putative transgenic Alpha potatoes

Line Description	% PVX infection	% PVY infection
6	1	20
7	8	25
27	6	4
30	15	3
32	20	0
33	10	0
34	0	3
36	0	8
37	10	10
38	10	8
39	10	0
42	5	0
44	20	0
45	8	15
46	3	0

* Percentages of infection of the 15 best lines tested at Prosser, Washington, U.S. Forty plants of each line were inoculated mechanically (PVX) and aphid-mediated transmission (PVY). At the end of the season, all plants were evaluated by ELISA, and the percentage of the infected plants calculated.

A second test was conducted in Mexico (Irapuato, Guanajuato) with some of the best lines from the Prosser test and some new untested lines (30 lines in total). The experimental protocol was similar to the one in Prosser. Some lines repeated the good performance observed in Washington state, whereas some of the new untested lines gave promising performances.

The second potato field test exposed CINVESTAV personnel to several important regulatory issues such as permit applications and selection of the location for the test. Although several field tests had been performed in Mexico by American companies, the potato test was actually the first one organized and performed by a Mexican institution.

PERSPECTIVES

Although improvement of Alpha variety certainly will have an impact on Mexican potato cultivation due to Alpha's wide acceptability, in some areas the impact will be predictably less noticeable. It is known that in some regions small farmers can not afford to cultivate Alpha because of its disease susceptibility when grown without several applications of chemical products. Thus some varieties, although lower in quality, have been adopted regionally, mainly by small farmers because they require a lower investment than Alpha. This observation, as well as the success obtained in Phases One and Two of the project, have convinced us to include other varieties in this effort. We expect that the overall impact of the project will be enhanced by increasing the options offered to small farmers.

The Mexican National Potato Program, highly motivated by Dr. John Niederhauser, has developed several new varieties, some of which have become very popular in Central America. However, in Mexico only a few have been adopted regionally, and no one has shown the potential to replace Alpha in the short-term. More recently however, there are indications that certain new varieties (some not even released yet) could become important in the long-term because they resemble Alpha in some agricultural aspects and present the advantage of being less susceptible to fungal diseases (e.g. *Phytophthora infestans*). These new varieties are the ideal candidates to be incorporated into the program for PVX and PVY resistance. The final decision is being made according to their response to the established tissue culture and transformation protocols as well as their potential impact for potato cultivation in Mexico.

ADDITIONAL IMPLICATIONS OF THE PROJECT

As mentioned above, the project also represented a vehicle to facilitate the establishment of regulatory procedures and the development of biosafety guidelines for testing and introduction of transgenic materials. This project, including the field test carried out in 1993 at Irapuato, Guanajuato, presented some new and interesting perspectives.

An informal Biosafety Committee was established in 1988-1989 to deal with some permit applications to import and field test transgenic material in Mexico by American companies. After some reorganization of the original group, an official Biosafety Committee was established as part of the Plant Health Administration of the Ministry of Agriculture and Water Resources (Dirección de Sanidad Vegetal, Secretaría de Agricultura y Recursos Hidráulicos). Since then, this committee has been evaluating all applications for transgenic field tests. All previous applications to test transgenic material in Mexico involved plants obtained by American companies in their U.S. laboratories. Those materials previously were field tested in the U.S., and by the time of the test in Mexico there was enough information about field performance and encountered or potential problems. An application submitted to the Biosafety Committee usually includes a USDA-APHIS Environmental Risk Assessment, and there is always an open communication between APHIS and the Mexican Biosafety Committee about the test.

This project presents some differences because some materials have been or will be tested in Mexico for the first time; thus there are no separate data about performance and possible risks (risk assessment). More importantly, the transgenic material is supposed to be released in massive amounts in the future and in places where wild potato relatives are known to be present. Therefore a new series of aspects have to be considered (cross pollination, etc.). In harmony with this project, ISAAA organized a series of activities oriented to expose the personnel involved in this project (from CINVESTAV) and the members of the Biosafety Committee to the experiences on regulatory issues for biosafety shared by members of agencies from the U.S., Canada and Europe.

As a summary, some of the expected impacts of this project can already be perceived. The phases on regulatory issues and biosafety as well as the training of Mexican scientists have been completed, and the results can jointly be evaluated with the transgenic potatoes already obtained in Mexico and their field-test data. The development of a product to be released to Mexican farmers has advanced according to the program and future tests will be required to select the best candidates. However, the most important impact, increasing potato production and decreasing production costs and environmental hazards by a reduction of chemical applications, will have to be evaluated in the long-term. This publication, by sharing our experience, can be considered as part of the last impact area where this project can be used as a model for other possible projects of technology transfer.

REFERENCES

1. Dirección General de Estadistica. Secretaría de Agricultura y Recursos Hidráulicos, 1992.
2. Beachy RN, Loesch-Fries S, Tumer N. Coat protein-mediated resistance against virus infection. Ann Rev Phytopathol 1990; 28:451-474.
3. Wilson TMA. Strategies to protect crop plants against viruses: Pathogen-derived resistance blossoms. Proc Natl Acad Sci USA 1993; 90:3134-3141.
4. Lawson C, Kaniewski W, Haley L et al. Engineering resistance to mixed virus infection in a commercial potato cultivar: resistance to potato virus X and potato virus Y in transgenic Russet Burbank. Bio/Technology 1990; 8:127-134.
5. Kaniewski W, Lawson C, Sammons B et al. Field resistance of transgenic Russet Burbank potato to effects of infection by potato virus X and Potato virus Y. Bio/Technology 1990; 8:750-754.

ASPECTS OF TECHNOLOGY TRANSFER TO BRAZIL:
APPROACHES USED BY THE BRAZILIAN AGRICULTURAL RESEARCH CORPORATION (EMBRAPA)

Maria José Amstalden Sampaio

INTRODUCTION

Mankind's great challenge over the next few decades will be to feed a growing population which may add another five billion inhabitants by the year 2025. Ninety percent of this population growth will occur in countries of Africa, Asia and Latin America. At present, there is a strong world tendency to seek sustainability. This sustainability is of both an ecological and socio-economic nature for agriculture, forestry and agroindustry.

EMBRAPA's aims, in harmony with world tendencies, address environmental quality, technology modernization of the agribusiness complex by means of territorial organization, farming production systems and sustainable management.[1] Within this context, technologies for the improvement of production and competitiveness are a strategic need. Thanks to efforts which started in 1982, EMBRAPA has grown to have good technological and scientific capabilities for agricultural biotechnology and has worked for plant biotechnology transfer in particular (see Table 12.1 for a list of EMBRAPA research centers).

Within the context of Art. 16 of the Convention on Biological Diversity[2] both access and transfer of technology related to the conservation and sustainable use of biological diversity among Contracting

Plant Biotechnology Transfer to Developing Countries,
edited by D.W. Altman and K.N. Watanabe. © 1995 R.G. Landes Company.

Table 12.1. EMBRAPA Decentralized Units

Acronym	English name	State Location	1994 Total Staff
a. Agroforestry or Agricultural Ecoregional Research Centers			
CPAA	Agroforestry Research Center for Western Amazonia	Amazonas	342
CPAC	Cerrados Agricultural Research Center	Districto Federal	506
CPAF-AC	ACRE Agroforestry Research Center	Acre	117
CPAF-AP	AMAPÁ Agroforestry Research Center	Amapá	76
CPAF-RO	RONDÔNIA Agroforestry Research Center	Rondônia	153
CPAF-RR	RORAIMA Agroforestry Research Center	Roraima	84
CPAMN	Center for Agricultural Research in the Mid-North	Piauí	301
CPAO	Center for Agricultural Research in the Mid-West	Mato Grosso do Sul	106
CPAP	Pantanal Agricultural Research Center	Mato Grosso do Sul	140
CPACT	Agricultural Research Center for Temperate Climate	Rio Grande do Sul	416
CPATC	Center for Agricultural Research of the Coastal Tablelands	Sergipe	202
CPATSA	Semi-Arid Agricultural Research Center	Pernambuco	450
CPATU	Agroforestry Research Center for Eastern Amazonia	Pará	630
CPPSE	Center for Research on Cattle Raising in the South-East	São Paulo	127
CPPSUL	Center for Research on Cattle Raising in Southern Fields	Rio Grande do Sul	122
b. National Commodity Centers			
CNPA	National Research Center for Oleaginous and Fibrous Plants	Paraíba	215
CNPAF	National Rice and Beans Research Center	Goiás	365
CNPC	National Goat Research Center	Ceará	161
CNPF	National Forestry Research Center	Paraná	155
CNPGC	National Beef Cattle Research Center	Mato Grosso do Sul	249
CNPGL	National Dairy Cattle Research Center	Minas Gerais	396
CNPH	National Vegetable Crop Research Center	Districto Federal	261
CNPMF	National Cassava and Tropical Fruit Research Center	Bahia	241
CNPMS	National Corn and Sorghum Research Center	Minas Gerais	414
CNPSA	National Pig and Poultry Research Center	Santa Catarina	202
CNPSO	National Soybean Research Center	Paraná	327
CNPT	National Wheat Research Center	Rio Grande do Sul	263
CNPUV	National Grape and Wine Research Center	Rio Grande do Sul	161
c. Basic Theme National Centers			
CENARGEN	National Genetic Resources and Biotechnology Research Center	Districto Federal	288
CNPAB	National Agro-Biology Research Center	Rio de Janeiro	142
CNPAT	National Research Center for Tropical Agroindustry	Ceará	133
CNPDIA	National Center for Research and Development of Agricultural Instrumentation	São Paulo	43
CNPMA	National Research Center for Monitoring and Assessment of Environmental Impact	São Paulo	122
CNPS	National Soil Research Center	Rio de Janeiro	188
CNPTIA	National Center for Technological Research on Information in Agriculture	São Paulo	42
CTAA	National Agroindustrial Food Technology Center	Rio de Janeiro	138

Parties are considered essential elements for the attainment of the objectives of the Convention. This has been one of the most difficult items during the discussions among nations in preparation for the conference which took place in Rio de Janeiro during 1992. However, modern biotechnology and mostly plant biotechnology, has been transferred, further developed and applied in developing countries for more than two decades.

EMBRAPA'S BIOTECHNOLOGY NATIONAL PROGRAM

POLICY AND ACHIEVEMENTS

Officially established as a National Program in 1989, the program aimed to optimize investment by avoiding duplication of efforts and to integrate researchers' desires for obtaining the best results from focused research with a relatively small number of trained staff. Based on the argument that any biotechnology project would need germplasm as a natural source of new genes, EMBRAPA decided to concentrate its major investment in capacity building and infrastructure at the National Research Center for Genetic Resources and Biotechnology (CENARGEN), its former Genetic Resources National Research Center. Different strategies have been used through the years to overcome major constraints, such as the high cost of the new biotechnologies being introduced within the framework of an organization with lower budgetary expectations, pressures exerted by managers for readily usable, short-term results, the need for an understanding of technical requirements and most importantly, the effective integration of biotechnology-based approaches with the conventional agricultural-research base.

Reviewed in 1993, during EMBRAPA's strategic planning, the program has now turned its focus to basic biotechnology and will give incentives to applied projects only if directly linked to a strong demand from the clients. Twenty four projects were submitted in 1994, and EMBRAPA allocated a budget of U.S. $1.48 million for laboratory supplies, small equipment and traveling expenses. The projects cover protein engineering, cloning and expression of new genes from microbes or plant wild relatives for resistance to pests and diseases, animal embryology and immunology, basic features of biological control, development of new applications for the gene gun, development of micropropagation for forest and fruit species, molecular characterization of N_2-fixing bacteria and domestic animals. Nine other projects that mainly involve the use of molecular markers were submitted to the other national programs.

CAPACITY BUILDING AND INFRASTRUCTURE

EMBRAPA, through the Ministry of Agriculture, receives the support of international loans from the World Bank (BIRD) and from the Interamerican Development Bank (IDB). Using these two agencies and national resources, EMBRAPA during the 1970s and early 1980s launched an aggressive human-resources upgrading program and sent more than 600 researchers abroad for their M.S. and/or Ph.D. degrees out of a total of 2,200 researchers. A similar number have been trained with grants from parallel programs run by the Ministry of Science and Technology through its agency, the National Research Council (CNPq). Out of this pool of leading research that can stimulate technology transfer, more than 150 scientists have developed their theses on subjects directly linked to biotechnology or have shifted from related areas to specialize in molecular or cellular biology.

Analyzed after so many years, mainly under the scope of biotechnology transfer, the program has been very successful because the great majority of these employees returned to Brazil and managed to stay with EMBRAPA to build up their teams. Of course, it is still too early to surmise that all potential problems are going to be solved. On the other hand, biotechnology itself is still under worldwide scrutiny to prove to society that applied science can bring about solutions which were promised more than 10 years ago. The continuous contact with former supervisors and the development of joint projects between CENARGEN and other EMBRAPA Research Units with various universities abroad have become a major technology transfer route. This exploration has and will continue to benefit both sides.

Under the same scope, another grant program was started several years ago by the Ministry of Science and Technology and has the acronym RHAE (Human Resources for Strategic Areas). RHAE aims to provide, on a competitive basis, grants for short- and long-term training abroad. Also there are provisions to fund grants for team-work, meaning that young M.Sc. researchers can be linked, to projects coordinated by senior investigators with advanced training. Both approaches have benefited EMBRAPA's researchers and technology transfer either by allowing more contact with laboratories in industrialized countries or by accelerating research projects through increasing competence from better-qualified, enthusiastic young researchers. The cost-efficient training of new scientists is also an important product of the program. Around 150 of these grants have been awarded to and implemented by EMBRAPA in the area of biotechnology since 1990.

Another mechanism utilized by EMBRAPA with both BIRD and IDB support which permits the rapid transfer of biotechnologies, has been the use of short-term consultants who have shown, in the vast majority of cases, an enormous interest in returning and/or in starting a collaborative program. As in many scientific environments, competition for external funding requires significant effort, which is time-consuming. Because consultants are invited within the framework of the Biotechnology Program, their projects usually answer a pre-selected and prioritized problem which has largely been documented by normal procedures. Therefore, this activity can catalyze the development of more comprehensive research proposals in an efficient and effective manner.

To facilitate the development of projects on several fronts, EMBRAPA has also invested in equipment and laboratory space. The last international project negotiated with IDB will allow seven research units located around the country to build special laboratories for biotechnological activities. CENARGEN should also complete construction in 1995 of a new 3,000 m^2 facility, partially financed by the World Bank, which will allow for the expansion of research and training activities. New state-of the art equipment has been acquired over time and has resulted in capacity for well-equipped laboratories. Investments in biotechnology-related equipment have reached more than U.S. $20 million for the facilities.

Major constraints are still the maintenance of imported equipment and fast acquisition of imported perishable reagents. Five important factors are responsible for the difficulties faced to give reasonable maintenance of imported sophisticated equipment: (a) the acquisition of spare parts is complicated and time-consuming, with high taxes for private firms; (b) experienced maintenance engineers able to work with all types of new equipment are not usually available; (c) when available in a few cases, maintenance services cost so much that the project budgets can not afford these allocations; (d) equipment is sold without the service manuals for fear of pirating, even for clients like EMBRAPA; and (e) there is lack of political decisions by EMBRAPA to build a maintenance network to assist the research units. Acquisition of reagents and expendable items also represents a serious bottleneck for the rapid development of science in the country. There are mechanisms that sometimes can bypass the difficulties, but for the bulk of reagents, the present regulations delay any importation by 8-10 months or even years, driving most researchers to despair. Companies such as Amersham, Pharmacia, Perkin Elmer, Biolabs and others are trying to minimize costs but the prices are still excessive (3-10 times the catalogue price) and impede research development.

EXTERNAL MECHANISMS
FOR TECHNOLOGY TRANSFER

At present, several international programs, specialized networks and institutions are operational and claim to direct their effort to technology transfer and capacity building. EMBRAPA is building up its own capacity to judge and react to this challenging new territory. The very fast changes in the world consensus about biosafety, intellectual property rights protection, plant breeders' rights, the Convention on Biological Diversity and international broad agreements such as GATT, MERCOSUR, NAFTA and others have overloaded researchers' and research managers' ability to keep up with aspects of technology transfer while continuing to conduct scientific research. Therefore, mechanisms which can simplify negotiations are to be highly considered at a time when the definition of protected subject matter regarding biological material, is still evolving and is far from fixed.

During the last two years, EMBRAPA has been in touch with ISAAA, the International Service for the Acquisition of Agri-Biotech Applications. At least three projects related to the development of Papaya Ring Spot Virus resistance in papaya, pest resistance in cotton and diagnostic tools for maize viral diseases have been brokered. The novelty behind the scheme is that ISAAA has as its specific mission the improvement of private-public collaborations.[3]

It is certainly too early to determine that this approach of using an intermediary will definitely be the way to build partnerships with industry. In terms of raising funds for projects, it seems to work very well for small countries and/or for crops that are not a significant factor for international commerce. For countries with a recognized well-developed agriculture such as Brazil, the negotiation of projects that involve traits for crops with high commercial potential has not been an easy task. The lack of an established legal framework on biosafety and some kind of intellectual property protection has caused substantial drawbacks. However, potential competition with industrialized nations has also been a major constraint.

INTERNATIONAL PARTNERS AND COLLABORATORS

Also helping to achieve EMBRAPA's goals in biotechnology transfer are partners such as the international centers belonging to the CGIAR system. CIAT, CIP, ICRISAT, ILRAD, IRRI, CIMMYT and IPGRI have worked together with CENARGEN and other Research Units for many years in the area of genetic resources. As biotechnology develops and the international centers begin to produce results themselves, EMBRAPA hopes that the same partnership can be enjoyed in this new field. Of course, it will all depend on how the CGIAR is going to put forward its policy regarding intellectual property.

An important agreement has been achieved with the ICGEB (UNIDO). Brazil has three Affiliated Centers recognized by the system: Oswaldo Cruz Institute (human health), Butantã Institute (vaccines) and CENARGEN (agricultural biotechnology). Training has been provided to several Brazilian scientists during the courses offered by ICGEB, and grants have been made available, on a competitive basis, to provide seed money for projects and support for international exchange.

MULTINATIONAL CORPORATIONS

Answering to an opening of Brazilian external policies during the last three years and with a possible stabilization of the economy, multinational corporations are showing an increasing interest in scientific joint ventures (Table 12.2). Some of these have been beneficial to both sides, with few worries about intellectual property rights. When the project involves the production of a GMO however, biosafety regulations must be addressed. An example case of this is herbicide resistance for crops such maize and soybean.

Numerous proposals have been put forth by companies such as Merck, Dow Elanco, Novo Nordisk, Monsanto and Ciba-Geigy and wait to be analyzed. The analyses should be done without preconceived ideas for the best mutual benefit. A growing interest is being generated in the area of biodiversity screening for different products such as new molecules for medical and cosmetic use, biological control and colorants. The country urgently needs to regulate access to its genetic resources, not to close frontiers, but to harmonize policies with neighboring countries and to have clear strategies for conducting negotiations with interested parties. A definite priority for technology transfer is the area of biochemical and molecular massive screening.

IICA AND PROCISUR

The Interamerican Institute for Agriculture Cooperation (IICA) has organized several programs, identified as PROCIs, that during the last ten years facilitated interactions between countries of a given Latin American region. The Southern Cone countries, Argentina, Bolivia, Brazil, Chile, Uruguay and Paraguay, through their national research organizations, belong to the program called PROCISUR, which has had a subprogram dedicated to biotechnology since 1992.

This mechanism has, up to now, provided a means for multilateral technology transfer. Recent strategic changes should allow this cooperative initiative to become a strong catalyst for acquisition of technology for the entire region. The collaboration will be innovative because it affords the potential for economies of scale for several laboratories, thereby making the countries more competitive to acquire a given technology that, if proprietary, will be negotiated only once for

---

Table 12.2. Examples of products and technology
to be negotiated with private industry

MONSANTO	Bt for cotton,
	Herbicide resistance for soybean
	Virus resistance for potatoes
	Bt for sugarcane
	Ripening control for tomatoes
ASGROW	Virus resistance for vegetable crops
CIBA-GEIGY	Bt and herbicide resistance for corn
ARACRUZ	Lignin gene for *Eucalyptus*

the block of countries. Training will also be negotiated for specific needs identified in several countries.

It is interesting to note that a possible conflict of interest may arise in the region when the free trade agreement (MERCOSUR) enters into full operation as of January 1995. Scientists hope to overcome that possible pitfall, at least for areas involving basic research.

LEGAL FRAMEWORK

Both from a scientific and judicial point of view, Brazil needs specific legislation for biosafety, plant protection and intellectual property rights. During 1993 and 1994, the political and scientific communities have been working on laws for these three major subjects, which would be in harmony with international trends in this area. The document drafts for plant variety protection and intellectual property rights are in Congress, waiting to be taken up and, if approved, implemented during 1995. The biosafety law was recently approved by Congress and will be in force beginning in 1995.

EMBRAPA has always defended the protection of intellectual property through a plant protection mechanism similar to that stipulated by the Union of the Protection of New Plant Varieties (UPOV) Convention in 1978. Brazil has not yet signed the International Undertaking on Plant Genetic Resources as proposed by FAO, but should do so if some modifications are accepted by other member nations. Therefore, this action could create a potential conflict with the recently-revised UPOV Convention of 1991. Moreover, patent protection for living organisms is a very sensitive issue and has complex ethical and legal implications, besides economic considerations.

The acceptance of very broad patents is another issue for serious discussion. For example, scientists worldwide are awaiting the resolution of the challenge for a patent awarded to Agracetus for rights to any form of transgenic cotton. This patent petition was filed in Brazil as early as December 1987. Up to now, the subjects of microorganism patenting and their appropriate definition have also been matters of great concern among the scientific community. There is a general consensus that plants, animals and other living organisms be excluded from this type of protection.

The first proposal for a biosafety law was sent to Congress as early as 1991, but only since May 1994 have the discussions been accelerated. The proposal was revised by scientists from the government and from private industry, and Congress granted final approval in December, 1994. Comprehensive guidelines should follow shortly after this approval, allowing for national and international projects to move towards conducting the first field tests of transgenic plants during 1995. Brazil has an enormous responsibility in the region, being one of the countries in the world rich in biodiversity and the center of diversity for several crops. All appropriate care will be taken to safeguard the ecosystems, the genetic resources and the health and well-being of citizens from any possible risks associated with the release of GMOs, and subsequently in the future with the commercialization of genetically-engineered food and other products.

REFERENCES

1. EMBRAPA. Environment and Development. 2nd Edition. Brasilia: SPI, 1993:67.
2. Convention on Biological Diversity. Rio de Janeiro: UNCET, 1992.
3. Raman KV. Facilitating Plant Biotechnology Transfer to Developing Countries. In: Altman DW, Watanabe KN, eds. Plant Biotechnology Transfer to Developing Countries. Austin, TX: RG Landes, 1995.

CHAPTER 13

USE OF PLANT BIOTECHNOLOGY TOOLS IN PLANT PROTECTION, GENETIC RESOURCES MANAGEMENT AND CROP GENETIC IMPROVEMENT

AN INTERDISCIPLINARY APPROACH WITH POTATOES AT THE INTERNATIONAL POTATO CENTER

Kazuo N. Watanabe, Jari P.T. Valkonen and Peter Gregory

INTRODUCTION

The recent developments in plant biotechnology are expected to have vast potential applications for solving bottlenecks and answering the current questions about productivity and sustainability in agriculture. This not only has implications for near-term applications in a short time frame, but also involves the employment of specific tools that can be achieved right now. In this chapter using the potato crop as an example, we would like to present some highlights of the biotechnology tools for solving pitfalls in: (1) plant quarantine, (2) plant genetic-resources management and seed program, and (3) crop genetic improvement.

Plant Biotechnology Transfer to Developing Countries,
edited by D.W. Altman and K.N. Watanabe. © 1995 R.G. Landes Company.

USE OF DNA PROBES
AND BIOCHEMICAL ASSAYS
FOR PLANT PEST DIAGNOSIS

Detection of virulent plant pests is an important issue in epidemiology to provide pest-free propagules for crop production at farmers' fields and for conservation of natural genetic resources in situ and at genebanks. This is especially important in developing countries, where farmers depend heavily on good quality propagules. Agricultural chemicals often can be cost-prohibitive and difficult to use for self-sustaining small farmers who are in the majority in many local communities.

Concerns encompass the spread of infection in production fields by pathogens as well as detection of materials contaminated by pests and potential disease agents. Thus, detection schemes should be operative for entire cropping systems, i.e. for plant quarantine inspection of "seed" programs and in growers' fields. Molecular probes can provide a powerful tool and augment serological tests with an accurate and efficient survey to monitor dangerous pathogens.

Potato spindle tuber viroid (PSTVd) is an unencapsidated single-stranded RNA molecule of 356-360 nucleotides, capable of independent replication in plant cells. PSTVd occurs in field-grown potatoes in North America, several South American countries, Russia and China.[1] It is considered an extremely deleterious pathogen for two reasons. First, it can reduce the yield of potato by 15-65%, and severe damage can also be caused in other solanaceous crops such as tomato and tobacco. Secondly, it is transmitted via tubers, pollen and true potato seed (TPS) and is, therefore, frequently detected in potato germplasm worldwide. Potato virus T (PVT) which occurs in South America is also a potentially dangerous pathogen and is transmitted through pollen and TPS.[2,3]

Most strains of PSTVd and PVT remain symptomless and may be present in low titers in infected potato plants. Therefore, detection based on the symptoms or the use of indicator plants is unreliable and time-consuming. While viruses can be tested serologically, no serological tests are available for the detection of PSTVd. Sensitive diagnostic tools such as the Nucleic Acid Spot Hybridization (NASH) test using DNA and RNA probes have been developed for rapid diagnosis of PSTVd in large numbers of samples.[4,5] The NASH test is carried out by partial purification of the RNA from the sample, spotting the sample on a carrier (usually a nitrocellulose membrane) and followed by hybridization with a specific RNA or DNA probe labeled with radioactive P^{32} or with nonradioactive "cold" labeling. DNA probes also have been developed for PVT, which has an RNA genome. This permits parallel detection of PSTVd and PVT once the RNA has been extracted from the samples, and facilitates large-scale processing.[4] DNA probes also have been developed for other potato RNA viruses,[6] but these probes still have their major impact in research and not in routine testing for plant quarantine purposes. The simplicity and reliability of the various serological tests available for other virus detection might make this conventional technology transfer an easier short-term solution. Protocols and methods for plant virus diagnosis are extensively reviewed in Matthews.[2]

The International Potato Center (CIP) supports a multilateral network on PSTVd quarantine. This network is conducted through an interdisciplinary team which consists of scientific officers and seed growers in the client developing countries and international scientists from different disciplines. Similar work is also conducted by the potato interdisciplinary team at Cornell University involving USDA and New York state scientists for the assistance of local potato growers with potential PSTVd problems and possible movement of the pathogen. International transfer of healthy germplasm also is enforced by such a system.

Potato ring rot disease is caused by the bacterium *Clavibacter michiganiensis* subsp. *sepedonicus* (Spieck & Kotth.) Skapt. & Burkh. It is a serious disease problem in North America and can cause a total loss of the potato crop if potato "seeds" are infected. The practice of cutting the seed tubers is an important means of spreading the bacteria. A zero-tolerance level is set for potato ring rot bacterium in imported plant stocks in many countries.[7] The bacterium is difficult to isolate from potato tubers and grows slowly on artificial growth media. However, due to the zero-tolerance level, very sensitive and reliable tools are required for diagnosis. Serological tests are useful, but cross-reaction with other bacteria may occur. Hybridization with DNA probes has proven to be more specific than the serological tests[8] and therefore would be the method of choice for potato ring rot detection in plant quarantine.

Polymerase chain reaction (PCR)-based techniques for plant diagnosis utilize designed primer sequences to amplify pathogen-specific DNA sequences in sample extracts up to a visual level. The presence of the product is taken as an indication of the presence of the pathogen. An advantage of PCR over other DNA-based techniques is that no probes are needed for detection of the amplified product. With viruses and PSTVd, a preceding reverse-transcription step is required to obtain a DNA copy of the RNA genome.[9] PCR-based techniques are qualitative and extremely sensitive. Therefore, they will probably become increasingly important in plant quarantine for the detection of pathogens with low-tolerance levels. Furthermore, PCR methods require simpler procedures, equipment and supplies, and these technologies can be transferred rapidly into programs in developing countries for research and for plant quarantine. PCR also has applications for other biological constraints such as detection of toxic substances for food safety and low levels of insect infestations.

BIOTECHNOLOGY TOOLS IN PLANT GENETIC RESOURCES MANAGEMENT FOR CONSERVATION AND BIODIVERSITY AND PROPAGATION FOR PRODUCTION

ESTIMATION OF DIVERSITY AND CLASSIFICATION OF GENETIC RESOURCES

Since the UNCED Convention at Rio de Janeiro in 1992, conservation of biological diversity in natural resources has been a growing concern for many nations, and monitoring such biodiversity is an essential matter to keep an adequate catalog of genetic resources and their practical use in industry. Enhancing biodiversity-conservation activities is indispensable to developing countries as they often depend on primary natural resources for their industrial base in addition to concerns about environmental protection.[10] Biotechnology tools can greatly facilitate biodiversity conservation and genetic resources management in various aspects for lesser-developed countries as well as developed nations.[11]

An estimation of genetic diversity of plant species can be accomplished by several ways, the most accurate method being the employment of molecular markers, especially DNA probes.[11] A detailed technical overview is described in Watanabe.[12] The molecular markers based on DNA probes can elucidate uncertainty in natural biodiversity and the genetic variation kept in either in situ or ex situ genebanks. DNA-based markers would provide an accurate measurement to detect change or loss of specific genetic variation which may result in a change in diversity of a plant species, and subsequently the overall ecological diversity. Further, by estimating changes in allele frequencies, molecular markers could target conservation efforts for particular alleles of interest which may be diminishing rapidly. Unsolved questions in biosystematics are amenable to DNA-marker technology to find out relationships between the plant species.

Genebanks receive many accessions, some without complete passport and phenotypic data, so that duplicates often exist. Identification of duplicates is tedious work, but a very important activity to increase efficiency of maintaining a large number of accessions in a genebank. DNA fingerprinting with molecular markers can support such objectives in addition to providing accurate identification of commercial varieties or other germplasm subject to concerns about associated intellectual property rights.

The same evaluations could simultaneously provide knowledge on general genetic variation existing among and between groups of accessions using multivariate analysis. Then, a well-documented group of accessions can be characterized as a core collection which represents a known degree of general genetic variation in given populations. This work is essential for genebanks that must keep many accessions at the same time in long-term storage, including cryopreservation, while a genetically-representative working collection should be continuously ready to use for clients.

PRESERVATION IN VITRO AND MICROPROPAGATION FOR CROP PRODUCTION

Tissue culture can facilitate crop germplasm preservation and crop production. Specifically, meristem culture with heat treatment can create propagules free of diseases caused by viruses and viroids. These new materials also can be preserved free of pathogens and pests in vitro for genebank purposes.[10,11,13] Plantlets can be established in vitro and used for transporting germplasm.[10,11] Wenzel[13] describes systems which are commonly used.

Cryopreservation using vitrification processes would facilitate long-term maintenance in quantity for inactive accessions.[11,14] For example, to maintain 5,000 potato genotypes in tissue-culture conditions with occasional subcultures every year, the annual operation costs would be U.S. $30,000. In contrast, cryopreservation using a liquid nitrogen freezer could accommodate the same number of clones with an annual cost of about U.S. $3,000. This method not only reduces routine operational costs, but can also reduce space requirements for keeping in vitro collections with modest facilities. Thus, this approach can be more easily managed in many lesser-developed countries. Very high recovery rates are now reported after cryopreservation so that almost all potato genotypes can be recovered after low-temperature exposure with the vitrification procedure (Steponkus, Cornell Univ., personal communication).

Micropropagation is widely used for scaling up production of propagules of vegetatively-propagated crops.[11] Small-scale micro-propagation for commercial seed production has already been adapted for use by many lesser-developed countries in the case of potatoes such as the ASEAN nations, Vietnam, China, Mexico, Southern Cone nations in South America and North African countries.[4,15] While large-scale operation to produce millions of propagules requires a high level of quality control to prevent unacceptable contamination, commercial production of vegetative propagules can be spread and adapted quickly for low-profile operation at the local-village level.[16,17]

CONSTRAINTS FOR GENETIC RESOURCES MANAGEMENT AND INTERNATIONAL TECHNOLOGY TRANSFER

The major constraints for conservation of vegetatively-propagated genetic resources such as potato and its wild relatives are the length of time and amount of resources required to preserve the selected germplasm undergoing international transport. Many countries require quarantine of exotic germplasm to reduce introduction of deleterious diseases and pests which could be carried by the germplasm. Effective phytosanitation, especially eliminating virus diseases by low-profile in vitro techniques, can take up to a year and often cost more than U.S. $1,000 per clone. This limits the number of individual clones that can be placed into the pathogen- and pest-free tissue-culture conditions for germplasm preservation. International funds for routine work on germplasm management is dwindling, so that maintenance activities for germplasm management are becoming more burdensome both for international genebanks and national germplasm repositories.[10]

Another concern is that mutual understanding of what constitutes transfer of phytosanitary clonal germplasm does not exist among countries, whereas all individual programs employ similar techniques for quarantine procedures and tissue culture-based germplasm conservation. This often hampers rapid introduction and utilization of clonal germplasm to users in different countries, especially those with strict quarantine regulations such as in Chile, Kenya, Philippines, North America, Europe and Japan. Without any verifiable data, it may take a year, sometimes two years, to receive the clonal germplasm from overseas genebanks after the plant quarantine authority finally obtains the materials. In order to facilitate international transfer of clonal potato germplasm, mutual understanding among representative genebanks and national plant quarantine programs is essential to balance phytosanitory needs for poorly-documented germplasm.[18] Plant biotechnology applications could greatly assist these well-known difficulties for germplasm exchange and effective utilization.

Intellectual property rights associated with plant genetic resources are another concern for germplasm management and transfer of various genetic resources between countries.[19-21] This aspect is discussed in detail in other sections of this book.

ESTABLISHMENT OF MOLECULAR MARKER ASSISTED INTROGRESSION AND SELECTION SYSTEMS FOR CROP GENETIC IMPROVEMENT

Advancing generations in potato breeding is dependent on information from the screening and selection processes. Screening procedures often require multiple repeated trials for confirming the results. This requirement may delay the enhancement practices, because it usually takes three to five years to put wild genes into cultivated potatoes.[12] However, this time scale should be regarded as rapid in comparison with previous efforts during the last three decades.[12]

DNA markers such as RFLPs and RAPDs are very powerful tools for understanding the genetic arrangement of the plant genome.[12,22] These tools have provided extensive opportunities to further comprehend the complicated genetics of many plant species. These tools are very helpful especially for a perennial species with a long life cycle such as trees, or a polyploid one like potato. The modern cultivated potato is a tetraploid, out-crossing species which is not amenable to complete analysis with conventional genetic means.

Several potato molecular maps were generated mainly using RFLP markers.[12,22] Map positions of these markers are well-conserved compared with alignment of the markers on the tomato molecular map, except for five major inversions on separate chromosomes.[22] Thus, the tomato map information is applicable directly to the ongoing potato-mapping activities. These potato maps were integrated to achieve a comprehensive analysis of the potato genome with applications to potato germplasm enhancement and breeding.

Potato cyst nematodes could damage potato crops severely by lowering both yield and the commercial value. Genetic improvement of potato is an important factor for controlling this pest in an integrated pest management program. In addition, conventional breeding procedures require enormous labor, time and logistics. For example, a conventional screening for the pest requires several months; on the other hand, the molecular markers can provide more precise information within a week for a large-scale breeding population. A single gene-inherited trait for cyst nematode resistance (*H1* gene) was mapped to chromosome V, and the marker information applied to various potato genetic stocks for the selection of the resistant clones.[23] These markers require further characterization for simplification with a PCR-based assay. Downstream application is slowly disseminating to developing-country programs with assistance from international organizations such as CIP. This is an example of the short-term application of molecular markers as a selection tool in plant breeding.

An important advance has been made to identify chromosomal regions which confer quantitative traits for potato improvement. Naturally-occurring glandular trichomes exhibit high levels of resistance to many harmful insect species and are quantitatively inherited.[24] As an alternative to pesticide usage, incorporation of host-plant resistance such as glandular trichomes would be an asset in an integrated pest management program. This resistance mechanism can reduce technical problems associated with adoption of IPM schemes for small farmers in developing countries who require simpler implementation procedures. In addition, applicator hazards from pesticides and cost constraints can be avoided with genetic modifications such as glandular trichomes.

With conventional genetic methods, the glandular trichome trait was under-exploited due to the complicated nature of its inheritance. At Cornell University efforts are made to select quantitative traits such as the glandular trichomes, as well as tuberization traits, tuber dormancy, etc., simultaneously in breeding populations.[12,24] While more intensive molecular-genetic information is essential for an accurate detection and utilization of such quantitative traits, progress has been reported on marker-assisted selection.[12] The work has been conducted by an interdisciplinary team which consists of geneticists, breeders, biochemists, entomologists, physiologists and nematologists to get feedback from each discipline to improve the knowledge and methods. Also, scientists from developing countries participate in this activity to enable transfer of the techniques to applications for their circumstances. CIP assists these interdisciplinary and multilateral activities as a catalyst for encouraging collaboration between developed and developing countries.[4]

Wild relatives of cultivated potatoes provide many valuable genes conferring resistance to other disease and insect pests that do not exist in the cultivated genepool.[11,12] The historical problem in utilization of valuable wild species has been difficulty in accurately detecting introgression of the target genes of interest while simultaneously selecting for elimination of deleterious exotic genes to derive elite genotypes. Rapid progress has been shown by the authors using the marker-assisted introgression of some wild *Solanum* species which requires only 3-5 years compared with the 10-20 year time frame taken for previous conventional methods.[11,12] This methodology is emerging with ongoing mapping efforts as a feasible technique for developing countries with limited resources.

TRANSGENIC POTATOES: A CASE OF INTERNATIONAL PROPRIETARY TECHNOLOGY TRANSFER FROM THE PRIVATE SECTOR TO THE GENERAL PUBLIC IN DEVELOPING COUNTRIES

As an alternative method to incorporate pest resistance in potato, CIP formed a collaborative network with various institutions for genetic engineering. Most of the collaborative groups are universities and governmental programs, but the private sector also participates in collaborations with proprietary biotechnology for the benefit of developing countries, not principally for the profit of the companies.

CIP has been engaged in a collaborative research and development program with Plant Genetic System (PGS) of Belgium for transgenic potatoes with insect resistance since 1992. The genetic materials and technology provided from PGS are chimeric gene constructs made from *Bacillius thuringiensis* (Bt) δ-endotoxin genes and associated vectors (Peferoen, 1992).[25] The salient points of the collaboration are: (1) an involvement of proprietary technology transfer from the private sector to developing countries without significant profits back to the company; (2) concentration on a few target programs in the developing countries; (3) possibilities for expansion of the project to many developing countries interested in the proprietary technology transfer; and (4) the role of CIP as a catalyst for the technology transfer.

At an initial stage, national programs in developing countries, PGS and CIP exchanged visits, and scientists from national programs received adequate training regarding genetic engineering with PGS in Belgium. PGS also offered on-site training for field evaluation of transgenic potatoes by national programs. To achieve these objectives, CIP has arranged with the Belgian government for financial support for the project. While only operational costs for PGS are covered by the grant, benefits in biotechnology transfer exclusively were considered from the perspective of the developing countries. Furthermore, during the program, CIP encouraged the national programs to make comprehensive biosafety regulations along with collaborations by other international organizations like ISAAA. While CIP did not play a major role in the actual development of regulatory procedures, CIP has made a contribution as a catalyst by initiating the field testing of transgenic potatoes in developing countries.

ISAAA also plays a similar role as a catalyst for proprietary biotechnology transfer with potatoes, which is detailed by K.V. Raman in a separate chapter. The presence of broker organizations such as ISAAA is becoming important due to the complexity associated with proprietary technologics, thc nccd to continually monitor the impact of technology transfer, and for genuine technical issues.

REGIONAL NETWORKS AND COUNTRY-BASED PROJECTS AS EFFECTIVE MECHANISMS OF POTATO TECHNOLOGY TRANSFER

CIP assists regional networks globally. In Latin America, CIP established networks which address technology, information, training and genetic materials of potato.

PRACIPA (Programa Andino Cooperativo de Investigacíon en Papa) is operational for Andean countries such as Peru, Bolivia and Equador on potato production. PRECODEPA (Programa Regional Cooperativo de Papa) is organized for Central and Caribbean countries. PROCIPA (Programa Cooperativo de Investigacíon en Papa) is made for Southern cone countries in South America. CIP also assists with other agriculture and/or natural resources networks such as CONDESAN (Consortium for the Sustainable Development of the Andean Ecoregion), PROCISUR (Programa Cooperativo de Investigacíon Agricola del Cono Sur) and PICTIPAPA (International Collaborative Late Blight Program) in Mexico. Needless to say, CIP also facilitates technology transfer with other IARCs (International Agriculture Research Centers) in Latin America such as CIAT (Centro Internacional de Agricultura Tropical), CIMMYT (Centro Internacional de Mejoramiento de Maíz y Trigo) in Mexico and IICA (Instituto Interamericano de Cooperacíon para la Agricultura) in Costa Rica. Interactions with international NGOs such as CARE and TALPUY (Grupo de Investigacíon y Desarrollo de Ciencias y Tecnología Andina) are also formed to spread technology within local community levels. All of these networks and collaborative relationships with other international efforts also can facilitate potato biotechnology transfer in a multilateral approach. This not only provides the technology, but also a possible mechanism to give feedback for improvements that ultimately benefit end-users.

Local programs are also emphasized to focus on potato technology transfer: PROINPA (Proyecto de Investigacíon de la Papa) in Bolivia, FORTIPAPA (Fortalcimiento de la Investigacíon y Produccíon de Semilla de Papa) in Equador and SEINPA (Semilla e Investigacíon en Papa) in Peru. CIP also collaborates strongly with bilateral international aid organizations, in corresponding local programs in Latin American countries such as USAID (U.S. Agency for International Development), IDRC (International Development Research Centre) in Canada, JICA (Japan International Cooperation Agency), COTESU (Cooperacion Tecnica Suiza) of Switzerland, and GTZ (German Agency for Technical Cooperation).

CLOSING STATEMENT

The concept and the tools used by the potato interdisciplinary team, are applicable to many other plant species for on-going research and development activities. Plant biotechnology tools must be supported by a rather long-term recognition and understanding both by the public and private sectors. As always, new technology may need to have refinements through its implementation phase, like the marker technology for crop genetic improvement, while efforts can be focused for short-term applications and utilization. Mutual understanding and good will on the part of the technology donor/developer and the recipients are essential keys in making the outcome of research fruitful for the donors and users alike.

Newer technologies for creating GMOs by the insertion of transgenes are emerging, and biosafety and food-safety issues as well as IPR concerns must be examined carefully for each program in the client developing countries.[26,27] The safety of a technology, the acceptance by the public and users and the benefits from the outcomes of a technology should be well understood in the area of plant biotechnology before starting a large-scale application for the commercialization phase.[17,28] Genetic engineering may require a longer time frame to be applied for practical uses which obtain consensus understanding and agreement among regulators, biological and social scientists and growers as well as the general public. On the other hand, diagnostic probes, the molecular probe/marker technology and micropropagation should be regarded as widely and immediately applicable for regular use if cost effectiveness is demonstrated. However, further dissemination of the technology and widespread utilization are future tasks for both technology donors and users. Worldwide, multilateral partnerships would help make technology available for more users, which could lead to an effective technology network and the enhancement of productivity and sustainability by the technology.[29]

REFERENCES

1. Owens RA. Khurana SMP, Smith SR, Singh MH, Garg, ID. A new mild strain of potato spindle tuber viroid isolated from wild *Solanum* spp. in India. Plant Disease 1992; 76:527-529.
2. Matthews REF. Diagnosis of Plant Virus Diseases. London: Academic Press, 1993.

3. Salazar LF, Harrison, BD. Host range, purification and properties of potato virus T. Ann Appl Biol 1978; 89:223-235.
4. CIP. Annual Report. Lima: International Potato Center, 1992.
5. Salazar LF, Balbo I, Owens, RA. Comparison of four radioactive probes for the diagnosis of potato spindle tuber viroid by nucleic acid spot hybridization. Potato Res 1988; 31:431-442.
6. Baulcombe DC, Fernandez-Northcote EN. Detection of strains of potato virus X and a broad spectrum of potato virus Y isolates by nucleic acid spot hybridization (NASH). Plant Disease 1988; 72:307-309.
7. De Boer SH, Slack SA. Current status and prospects for detecting and controlling bacterial ring rot of potatoes in North America. Plant Disease 1984; 68:841-844.
8. Drennan JL, Westra AAG, Slack SA, Delserone LM, Colmer A, Gudmestad NC, Oleson AE. Comparison of a DNA hybridization probe and an enzyme-linked in field-grown potatoes. Plant Disease 1993; 77:1243-1247.
9. Hadidi A, Montasser MS, Levy L, Goth RW, Converse RH, Madkour MA, Skrzeckowski LJ. Detection of potato leaf roll and strawberry mild yellow-edge luteoviruses by reverse transcription-polymerase chain reaction amplification. Plant Disease 1993; 77:595-601.
10. Wilkes G. Strategies for Sustaining Crop Germplasm Preservation, Enhancement, and Use. Washington DC: Consultative Group on International Agriculture Research, 1992.
11. Dodds JH, Watanabe K. Plant genetic resources management and biotechnology. Diversity 1990; 6(3,4):26-28.
12. Watanabe K. Potato molecular genetics. In: Bradshaw J, MacKay G, eds. Potato Genetics, Wallingford: CAB International, 1994:213-235.
13. Wenzel G. Tissue culture. In: Bradshaw, J, MacKay G, eds. Potato Genetics, Wallingford: CAB International, 1994:173-195.
14. Steponkus P. Fundamental aspects of cryoinjury as related to cryopreservation of plant cells and organs. In: Zaitlin M, Day P, Hollaender A, eds. Biotechnology in Plant Science, New York: Academic Press, 1985:145-160.
15. CIP. Annual Report. Lima: International Potato Center, 1990.
16. ISAAA. Annual Report 1993. Ithaca, NY: The International Service for the Acquisition of Agri-Biotech Applications, 1993.
17. Pardey PG, Roseboom J, Anderson JR eds. Agricultural Research Policy—International Quantitative Perspectives. New York: Cambridge University Press, 1989.
18. Watanabe K. Plant genetic resources in Andes. Kagaku 1991; 61(11):717-718 (Japanese).
19. Baenziger PS, Kleese RA, Barnes RF eds. Intellectual Property Rights: Protection of Plant Materials. Madison: Crop Science Society of America, 1993.
20. Barton JH, Siebeck WE. Intellectual Property Issues for the International Agriculture Research Centers. Washington DC: Consultative Group on International Agriculture Research, 1992.

21. Lesser W. Equitable Patent Protection in the Developing World, Issues & Approaches. Christchurch: Eubios Ethics Institute, 1991.
22. Tanksley SD, Ganal MW, Prince JP, de Vicente MC, Bonierbale MW, Broun P, Fulton TM, Giovannoni JJ, Grandillo S, Martin GB, Messeguer R, Miller JC, Miller L, Paterson AH, Pineda O, Röder MS, Wing RA, Wu W, Young ND. High density molecular maps of the tomato and potato genomes. Genetics 1992; 132:1141-1160.
23. Pineda O, Bonierbale MW, Plaisted RL, Tanksley SD. Identification of RFLP markers linked to the H1 gene coferring resistance to the potato cyst nematode (*Globodera rostochiensis*). Genome 1992; 36:152-156.
24. Bonierbale MW, Plaisted RL, Pineda O, Tanksley SD. QTL analysis of trichome-mediated insect resistance in potato Theor Appl Genet 1994:973-987
25. Peferoen M. Engineering of insect-resistant plants with bacillius thurigiensis crystal protein genes. In: Gatehouse AMR, Hilder VA, Boulter D. eds. Plant Genetic Manipulation for Crop Protection, Wallingford: CAB International, 1992:135-154.
26. Lesser W, Maloney AP. Biosafety: A report on regulatory approaches for the deliberate release of genetically-engineered organisms. Ithaca, NY: Cornell International Institute for Food, Agriculture and Development, 1993.
27. Persley GJ, Giddings LV, Juma C. Biosafety. The safe application of biotechnology in agriculture and the environment. The Hague: International Service for National Agriculture Research, 1992.
28. Kung SD, Wu R eds. Transgenic Plants Vol. 2. Present Status and Social and Economic Impacts. New York: Academic Press, 1993:191-246.
29. Tribe ED. Doing Well by Doing Good. Agricultural Research: Feeding and Greening the World. Leichhardt: Pluto Press, 1991.

REVIEWS FROM ORGANIZATIONS FACILITATING PLANT BIOTECHNOLOGY TRANSFER

THE ROCKEFELLER FOUNDATION'S INTERNATIONAL PROGRAM ON RICE BIOTECHNOLOGY

Gary H. Toenniessen

INTRODUCTION

The Rockefeller Foundation's International Program on Rice Biotechnology is an integrated set of research, training, technology transfer and capacity-building activities structured to produce improved rice varieties that will benefit low-income rice producers and consumers in developing countries. The individual grants or fellowships supported under the Program address one, and usually more than one, of the following objectives:

1. to assure that the techniques of biotechnology are developed for tropical rice;
2. to create sufficient biotechnology capacity in rice-dependent countries to meet current and future challenges to rice production;
3. to better understand the consequences of agricultural technical change in Asia, in part to help in setting priorities for biotechnological applications; and
4. to apply this knowledge and capacity to the production of improved seed and other materials used by farmers.

The Program was first approved by the Foundation's Trustees in December 1984 with the realization it would evolve over time and

Plant Biotechnology Transfer to Developing Countries,
edited by D.W. Altman and K.N. Watanabe. © 1995 R.G. Landes Company.

require a long-term commitment. Initially, funds were concentrated on generating enabling technologies such as molecular genetic maps of rice and of major pathogens of rice, gene tagging and marker-aided selection and genetic engineering. Priorities were established for applications of rice biotechnology in Asia as a whole and refined for individual countries. As the new tools of rice biotechnology became available, greater emphasis was placed on training, technology transfer, capacity building and research on priority traits. Such activities now receive a majority of Program funding. Within the next few years the use of biotechnology should make it possible to produce new rice varieties possessing useful characteristics that could not have been produced using conventional plant breeding alone.

This chapter describes the origins of the International Program on Rice Biotechnology, its evolution over the past ten years, its progress to date and its anticipated results.

ORIGINS OF THE PROGRAM

Rockefeller Foundation funding for international agricultural research dates back to 1943 with the establishment of the cooperative Mexican Agricultural Project. As part of this project, high-yielding semi-dwarf wheat varieties were developed which performed well in India and Pakistan as well as Mexico. This demonstrated that improved seed could be an effective mechanism for translating scientific progress into benefits for poor people over broad areas. To apply this same strategy to rice the Rockefeller Foundation joined with the Ford Foundation in establishing the International Rice Research Institute (IRRI) in the Philippines. This was the first of what are now 17 international agricultural research centers supported by over 40 donors, in aggregate known as the Consultative Group for International Agricultural Research (CGIAR). In addition to IRRI, the International Center for Tropical Agriculture (CIAT) in Colombia and the West African Rice Development Association (WARDA) in the Ivory Coast have responsibility for working with national programs to produce improved rice varieties.

These international centers are key players in what has become a highly effective international rice research system. The centers draw upon discoveries made and technologies generated in advanced laboratories worldwide. They conduct strategic and applied research and deliver the results, in the form of improved breeding lines, appropriate agronomic practices and relevant new knowledge, to collaborating national rice improvement programs where new varieties are adapted to local conditions and released to farmers.

The modern rice varieties and management practices generated by this research system have contributed significantly to increased productivity, greater food self-sufficiency and economic development in numerous countries. They were particularly effective in Asia where farmers for centuries have used irrigation, organic fertilizer and hand weeding on their small holdings. These farmers readily adopt improved varieties and, using their traditional intensive-management practices combined with purchased inputs, have in some places pushed yields close to their full potential. By the late 1970s over half of all Asian rice-growing land was planted with these modern varieties, benefiting literally billions of people who daily consumed the lower-cost end product. Throughout the 1970s per capita grain production in Asia increased significantly and over the past 20 years, the proportion of the population affected by under-nutrition declined from roughly 40% to 20%.[1]

While these were significant accomplishments and encouraging trends, beginning in the 1980s there were disturbing indications that rice scientists and rice farmers in Asia were pushing existing technology about as far as it could go. The increases in rice production which characterized the 1960s and 1970s began to slow, and yields began to plateau in some countries. An additional threat began to materialize in some locations that farmers would irreparably damage the resource base as they sought to feed more and more people.

Yet, the countries of South and Southeast Asia were still experiencing rapid population growth primarily due to a decline in the death rate, caused in part by improved nutrition. An expected concomitant decline in fertility rates was also occurring, but the momentum built into the population structure meant that the number of people living in the rice-dependent countries of Asia would surely more than double. It was feared that agriculture did not have the technology to achieve the further doubling or tripling of food production necessary for these countries to reach a stable population fed by sustainable agricultural systems. New yield-enhancing and resource-conserving technologies would certainly be needed.

Scientists working in the emerging and still very unproven field of biotechnology were promising to provide just such technologies, but there were also many skeptics, particularly within the agricultural research establishment. Following a major review of the Foundation's funding in international agriculture, the decision was made in 1984 to commit roughly half of the agricultural budget to a new program focused on rice biotechnology. The rationale for the Program as presented to the Foundation's Board of Trustees can be summarized as follows:

1. In terms of production and consumption, rice is by far the most important crop in the developing world accounting for up to 80% of the calories consumed in some of the poorest countries of Asia. Moreover, projections indicated that in the 1990s and beyond, in many developing countries, growth in demand for rice would move ahead of supply.

2. Rice breeding programs were mature at the international centers, such as IRRI, and in several Asian countries. If

new breeding tools could be developed, an international rice research system was already in place which could put them to effective use and deliver improved seed to farmers.

3. Rice was a highly neglected crop with regard to biotechnology. A survey conducted in 1983 did not find a single research program in the U.S., Europe or Australia studying the molecular genetics of rice. This was in sharp contrast to maize, potatoes and wheat, on which considerable research was already under way in industrialized countries.

4. It was realized that a biotechnology program focused on rice would require considerable investment of resources and staff effort over a long period, with no assurances that the science would generate results applicable to the practical objective of further increasing rice production. Neither the international centers nor most donors, other than a foundation, would have the flexibility to make a long-term investment in such a high-risk undertaking.

In December 1984 the Foundation's Board of Trustees approved the new Program and made an initial commitment of over three million dollars. By December 1994 the Foundation's investment in rice biotechnology totaled over $62 million as summarized in Table 14.1.

DEVELOPMENT AND EVOLUTION OF THE PROGRAM

CREATING RICE BIOTECHNOLOGY

The original plan for the Program was to begin with grants to advanced research laboratories enabling them to build a technical base of knowledge required to create rice biotechnology, and then to link these laboratories to international centers and national institutions in Asia responsible for rice improvement. Most of the world's leading

Table 14.1. Summary of the Rockefeller Foundation's Financial Investment in the International Program on Rice Biotechnology ($ millions)

	Total 1984 - 1994	1994 only
Advanced biological research	25.0	2.0
RDC[a] research	13.4	2.4
RDC fellowships	12.6	1.6
International centers	7.5	1.0
Social science research	2.8	0.3
Networking and information dissemination	1.4	0.4
Total	62.7	7.7

[a]RDC - Rice-dependent country

plant molecular biology laboratories were contacted about the Program, and a formal request for proposals was widely distributed and published in leading scientific journals. The proposals received were sent out for peer review and in some cases site visits with review teams were employed. The result was the selection in 1985 and 1986 of 15 laboratories in the U.S., Europe and Japan that were at the forefront of plant molecular and cellular biology. They were funded to generate the knowledge base and scientific tools necessary to create the applied science of rice biotechnology. With the exception of the two laboratories in Japan, none had done significant prior work with rice.

In 1986, Dr. Robert Herdt, an economist, joined the Foundation's staff to establish priorities for the Program. These took the form of desired traits, listed in Table 14.2, and served as specific targets for the Program during the initial years.[2] A series of workshops were then held designed to assess alternative research strategies for achieving the priority traits. Participants included rice scientists with experience relevant to the trait of interest (e.g. blast resistance and drought tolerance), molecular biologists at the forefront of biotechnology research on the trait (usually in plant systems other than rice), scientists already funded to develop the tools of rice biotechnology and knowledgeable scientists who were not potential grantees and who served as advisors to the Foundation. While the official product of each workshop was a report recommending one or more promising research strategies, the unofficial product was a new set of research proposals which reflected these recommendations and involved collaborations amongst participants in the workshop. With input from the advisors, a subset of these proposals was selected for funding.

In this way, the Foundation was able to add to the Program biotechnology laboratories at the forefront of work on particular traits and to link them to scientists knowledgeable about that trait in rice and to laboratories already working on development of the enabling technologies. All 46 industrial-country laboratories currently in the Program are working on such collaborative projects. Their progress, renewal

Table 14.2 Priority traits for rice biotechnology research in Asia [2]

Resistance to tungro virus
Submergence tolerance
Resistance to gall midge
Male sterility
Resistance to brown planthopper
Lodging resistance
Seedling vigor
Tolerance of water logging
Resistance to drought/blast complex
Resistance to yellow stemborer
Resistance to ragged stunt virus
Cold tolerance
Drought tolerance
Resistance to leaf folder
Weed tolerance/control
Resistance to sheath blight
Apomixis
Resistance to bird damage
Resistance to storage insects
Resistance to bacterial blight
Resistance to blast

proposals and contributions to training and technology transfer are evaluated every two or three years by the Program's Scientific Advisory Committee and Foundation staff. The prospect of renewed funding has been a key factor in keeping these laboratories committed to the goals of the Program. A portion of each grant is now designated for technology transfer. With supplemental funding several have offered formal training courses for developing-country scientists and nearly all have hosted Foundation-sponsored fellows. These same laboratories have assumed leadership of subnetworks within the Program that are focused on particular traits such as durable blast resistance or use of particular technologies such as marker-aided selection.

THE ROLE OF INTERNATIONAL CENTERS

As noted previously the International Rice Research Institute in the Philippines and the International Center for Tropical Agriculture in Colombia have responsibility for and a significant record of accomplishment in producing improved rice varieties and in helping to build rice research capacity in developing countries. From the beginning it was expected that they would function as key actors in the International Program on Rice Biotechnology. In 1985, however, they had limited capacity in biotechnology as was the case with most agricultural research institutions. Both had some capability in rice tissue culture, particularly anther culture, and IRRI was using embryo-rescue techniques as a means of producing wide hybrids, i.e. crosses between wild relatives of rice and cultivated rice. Neither had significant molecular biology capability, and they were not in a position to undertake work on the development of new molecular technologies.

In the late 1980s Program funding to IRRI and CIAT enabled them to strengthen their research in anther culture and wide hybridization, to begin transferring these technologies to national programs and to become part of a larger network working on development of more advanced biotechnologies. From 1986 to 1988 the Foundation seconded a cytogeneticist, Dr. Lesley Sitch, to work at IRRI on wide hybridization and to gather advice on further development of the Program from knowledgeable rice scientists. The first and second annual meetings of the Program were hosted by IRRI. This helped the centers to develop partnerships with other laboratories in the Program, and over time funding to the centers gradually shifted to research on application of new molecular tools. For a number of years IRRI used a portion of its Program funding to employ a shuttle scientist who moved back and forth between IRRI and Cornell facilitating the transfer of materials and technology in both directions. These types of partnerships greatly speeded the development and application of new technology and have contributed much to the success of the Program.

By 1990 it was evident that biotechnology would significantly impact plant breeding, and IRRI and CIAT began using their core funds to hire senior staff with molecular biology expertise. While they now have well-trained staff who can apply and further develop most of the tools of rice biotechnology, they continue to draw upon partners elsewhere as the source of inputs such as additional and more effective genetic markers, potentially-useful cloned genes and sophisticated strategies for controlling gene expression. With Program support the international centers have offered a series of training sessions for colleagues from national programs in Asia and Latin America. Recently IRRI obtained funding from other donors to expand significantly its research and training in rice biotechnology.

The West African Rice Development Association was reorganized in the late 1980s. While relatively small, it has become an effective CGIAR-funded center serving Africa. From new headquarters in the Ivory Coast, WARDA has established productive rice-breeding programs and has built excellent linkages with national programs in its region. Program grants to WARDA are intended to strengthen its rice anther-culture research, to provide staff training, and to lay the foundations for future application of more advanced biotechnologies. Much as CIAT is serving as a conduit for delivery of new technologies to national programs in Latin America, WARDA should soon be serving as a conduit for delivery of new technology to national programs in Africa.

During the past five years a variety of new international organizations have come into existence dedicated to the safe transfer of biotechnologies to developing countries. While their track records and budgets are not equivalent to that of IRRI and CIAT, they are addressing important problems and making useful contributions, particularly with regard to capacity building. The following contribute to the International Program on Rice Biotechnology.

The International Center for Genetic Engineering and Biotechnology (ICGEB) was established by the United Nations Industrial Development Organization with agricultural biotechnology research and training facilities in New Delhi, India. Program support has enabled ICGEB to strengthen its research on gene tagging and marker-aided selection in rice breeding, and to offer a series of biotechnology training programs. ICGEB scientists travel to and provide hands-on assistance and training to collaborating national programs in South Asia.

The Center for the Application of Molecular Biology to International Agriculture (CAMBIA) headquartered in Canberra, Australia is a newly-established research and training organization committed to producing inexpensive biotechnology tools which can be effectively utilized in developing countries. Program support has enabled CAMBIA to produce a "rice transformation kit and protocol" which CAMBIA scientists hand-deliver to national programs. Follow-up training and assistance are then provided.

The International Service for Acquisition of Agri-Biotech Applications (ISAAA) is an international organization committed to the acquisition and transfer of proprietary agricultural biotechnologies from the industrial countries for the benefit of the developing world. Program support to ISAAA's AsiaCenter in Japan is intended to facilitate the transfer of proprietary rice biotechnologies to rice-dependent countries of Asia. ISAAA also organizes regional biosafety workshops, including one focused on rice which was held in Indonesia.

STRENGTHENING NATIONAL PROGRAMS

One of the principal goals of the Program is to help build a comprehensive capability in rice biotechnology in the rice-dependent countries of Asia. This means having the indigenous scientific capacity to apply biotechnology in rice breeding, to solve problems when they occur, to utilize effectively further advances in the science and to create new and improved technology. In addressing this objective, support is provided to a spectrum of national research institutions, including universities and fundamental research institutes, as well as those institutes responsible for rice breeding. In several countries, the Program is implemented in collaboration with national agencies that provide leadership, coordination and local-currency funding.

China was the first developing country to join the Program and serves somewhat as a model. In 1985 grants for research in rice anther culture and training in the use of molecular technologies were awarded to the China National Rice Research Institute and the Shanghai Academy of Agricultural Sciences. Shortly thereafter a "memorandum of understanding" was signed with the China National Center for Biotechnology Development (CNCBD), a funding agency within the State Science and Technology Commission, establishing a "Cooperative Program on Biotechnology Development for Rice Improvement in China." A nationwide program of competitive research grants in rice biotechnology was announced, and over 50 proposals were received. An eight-person review team made up of consultants and staff from the Foundation and staff from CNCBD reviewed the proposals and conducted site visits in Beijing, Shanghai, Guangzhou, Hangzhou and Wuhan. Twenty projects were selected to receive research and training support.

Various training opportunities were provided for the principal investigators and their staffs. Over time the molecular biology research capability of these scientists was enhanced, and other highly capable scientists joined the Program following their return to China from training abroad. CNCBD has organized a series of meetings of the Chinese Program. Chinese-speaking consultants participate and advise the Foundation on the progress of each project, help CNCBD organize peer review and evaluation processes and interview fellowship candidates. Currently 34 research projects at 18 institutions are being supported as part of the Cooperative Program. All involve international collaborations, and most have published results in international peer-reviewed journals.

In 1988 Dr. John O'Toole was appointed a Foundation Field Staff Scientist, stationed initially in New Delhi and later in Bangkok, and charged with expanding the Program into South and Southeast Asia. In India a comprehensive program very similar to the one in China was developed. Administrative leadership and local funding are provided by the Indian Council for Agricultural Research and the Department of Biotechnology. The Directorate of Rice Research in Hyderabad organizes and usually hosts the annual meetings.

In Thailand a smaller but comprehensive program has been developed in cooperation with the National Center for Genetic Engineering and Biotechnology. This includes the development of rice biotechnology training programs at Kasetsart University designed to serve scientists from throughout Southeast Asia.

In other countries it is usually the national rice research institute that assumes leadership responsibility and coordinates several research projects located at various institutions. Sometimes it has been necessary to invest in the training of staff before biotechnology research could be undertaken.

By 1994 there were 76 developing-country research institutions participating in the Program with many receiving support for more than one research project. All of these institutions are eligible to send staff to the numerous training courses supported by the Program and to nominate staff for any of six types of fellowships listed in Table 14.3. As of September 1994 a total of 162 rice biotechnology fellows

Table 14.3. Fellowships awarded by International Program on Rice Biotechnology

Type	Description
Ph.D. Degree	Fellow receives Ph.D. degree training at advanced laboratory for 4 - 5 years.
Dissertation Research	Fellow receives Ph.D. degree in home country with a portion of dissertation research conducted at advanced laboratory for 1 - 2 years.
Postdoctoral Research	Fellow does postdoctoral research at advanced laboratory for 2 - 3 years.
Visiting Scientist	Fellow is senior scientist serving as visiting researcher at advanced laboratory or 1 - 2 years.
Biotechnology Career	Fellow conducts collaborative research requiring work at advanced laboratory for 3 months/year over a minimum of 3 years.
Technology Transfer	Fellow from advanced laboratory conducts collaborative research including work in a rice-dependent country for 3 months/year over 3 years.

had been selected. Nearly all receive training in and contribute to rice biotechnology research at one of the advanced laboratories in the Program, and after returning home they are provided with opportunities to maintain linkages with that laboratory.

Special mention should be made of the Biotechnology Career Fellowships. These fellowships are awarded to highly-capable young scientists from developing countries who returned home following a productive research experience abroad. The fellowships allow them to return three months a year, for a minimum of three consecutive years, to an advanced laboratory where they previously worked. They undertake a collaborative research project designed to utilize the strengths of both their home and host institutions. Thirty-eight Career Fellows have been supported as part of the International Program on Rice Biotechnology, some for nine years. All have also become the principal investigator for a Program-supported research project at their home institution, and nearly all collaborate with a host laboratory also supported by the Program. The Career Fellows significantly strengthen the linkages between their home and host institutions, making research at both more relevant to the goals of the Program. They become effective agents of technology transfer, literally carrying some of the most recent advances in rice biotechnology back to their home country every year. These fellowships are relatively inexpensive and one of the best investments made within the Program.

SETTING PRIORITIES AND ASSESSING IMPACTS

Support for social science research has been an important component of the Program from the beginning. As indicated previously, Dr. Robert Herdt, an economist who had spent ten years at IRRI assessing constraints to rice production and developing research priorities, took on the assignment of developing an initial set of research priorities for the overall Program. Using his prior experience plus inputs from many rice scientists and biotechnologists throughout the world, he made estimates of the potential value of desired traits and the probability that research using biotechnology would enable successful breeding for such traits. Traits that would especially help poorer farmers and/or be resource-conserving were given extra weighting in the priority calculations. The result was a list of priority traits (Table 14.2) which guided the Program during the initial years.[2]

These, however, were priorities for Asia as a whole. To make the results more useful at the national and local level, social scientists in each Asian country were supported to refine the method and to make priority calculations based on more detailed local information and national objectives. Social scientists at IRRI have also continued to refine the method to update and expand the information base and to revise priorities accordingly. Changes have occurred with tolerance to abiotic stresses now being considered more important and more achievable via biotechnology.

To better understand the likely impact of applying rice biotechnology in Asia, research was also supported to assess the differential economic and social effects, over time, of prior introductions of new rice production technologies on various populations in several Asian countries. The results were recently presented in the book *Modern Rice Technology and Income Distribution in Asia* which included a summary of the policy implications of alternative research strategies.[3]

NETWORKING

As part of the Program opportunities for networking and information sharing are provided. The *Rice Biotechnology Quarterly* attempts to report on all rice biotechnology research activities including people and events. A Rice Biotechnology E-Mail Network (RBNet) operates from Ohio State University. It facilitates communications and allows requests for information to be sent across the network. At Cornell University the RiceGenes Gopher Server allows for Internet connections to the United States Department of Agriculture (USDA)-funded RiceGenes Database. Through it, genetic maps and all available mapping data are available for downloading. Books summarizing the results of research, and consisting primarily of chapters written by Program participants, are occasionally published and widely distributed. Numerous other publications, dissertations, final reports and patent applications are sent to researchers working on relevant topics.

The 1994 annual meeting of the Program, hosted by Indonesia's Central Research Institute for Food Crops, in May 1994, brought together over 350 scientists. It included subnetwork meetings, special workshops organized by participants and presentations of everyone's research results. The Program's Scientific Advisory Committee and Foundation staff use the meeting to assess overall progress and to provide direction to the group as a whole. A variety of other meetings, including annual meetings of national programs and ad hoc meetings called to address specific problems or issues, are also supported.

These networking functions are, for the most part, available to anyone working in rice biotechnology, not just those currently receiving Program support. In most cases, Program participants remain part of the network even after their Foundation-funded projects are complete. Many have been able to obtain funds from other sources to continue their rice biotechnology research.

SCIENTIFIC PROGRESS

TISSUE CULTURE

The ability to grow plant cells, tissues and organs in artificial media and, following useful genetic manipulations, to regenerate whole plants is an important aspect of biotechnology. There were many developing-country scientists skilled in this technology at the time the Program was initiated. The early applications of biotechnology in rice research institutions were built upon this expertise. Anther culture, which can be used to produce doubled-haploids and thus speed the evaluation phase of breeding, and embryo-rescue techniques which allow breeding lines to be crossed with wild relatives of rice, have been particularly useful. New varieties produced utilizing these techniques and containing valuable traits, such as resistance to brown planthopper, resistance to bacterial blight and tolerance to cold, have reached farmers' fields and are performing well. Good progress is being made in expanding this technology to a broader range of varieties, particularly the tropical (indica) rices.

RICE GENOME MAPS AND MARKERS

Among crop plants rice (*Oryza sativa*) has turned out to have attributes which make it especially amenable to genome research. It has a DNA content smaller than any other grain species.[4] Rice is a true diploid with one copy of most genes per haploid genome and a relatively high percentage (ca. 75%) of single-copy DNA. This has made it a model cereal for cloning genes known by phenotype only.

A molecular genetic map of rice, consisting of over 700 markers was developed at Cornell University for use in rice genetics and breeding.[5] Repositories of mapped clones are maintained at Cornell, IRRI, CIAT, the Institute of Genetics in Beijing and the University of Georgia. These materials are available free-of-charge to scientists who request them. A kit containing the most recent version of the map, a set of over 100 reference clones and protocols for using markers has been distributed throughout the world. Libraries of bacterial artificial chromosomes containing large inserts of rice DNA were produced by and are available from the University of California at Davis[6] and Texas A&M University.[7] This should greatly facilitate physical mapping of the rice genome and gene cloning.

Molecular markers are now routinely used to study rice genetic diversity, classification and phylogeny for purposes of germplasm management.[8] At the 1994 rice biotechnology meeting in Indonesia scientists from over a dozen countries reported on their progress in finding markers tightly linked to rice genes of agronomic or economic interest and on using the markers to follow inheritance of the genes in breeding programs. Scientists are using the markers to pyramid multiple genes for resistance into elite breeding lines. Those lines containing combinations of resistance genes show greater resistance than lines carrying a single gene.[9] In some cases combination of resistance genes provide broader spectra of resistance than might be expected with simple additive gene action (Yoshimura et al, personal communication). The durability of the resistance is now being tested. Eventually IRRI hopes to provide breeders with "superdonor" lines which will allow multiple genes for resistance to be selected in a single cross.[10]

For a few quantitative traits of rice that are relatively easy to score, such as cooked-grain elongation, markers which detect quantitative trait loci (QTLs) have been identified.[11,12] For more complex traits, like drought tolerance, crosses which give significant genetic variation for component traits (e.g. osmotic adjustment, epidermal conductance, root/shoot ratio, root length, root thickness and root penetration) are being used for QTL analysis.[13-15] Two large research projects are under way in China using markers to understand better the genetic basis of heterosis in rice and to exploit it more effectively in hybrid rice production.[16]

Research is continuing aimed at developing rice molecular genetic-mapping technology into an inexpensive practical breeding tool. The ends of the most useful markers are being sequenced at IRRI to facilitate their distribution as data sets rather than bacterial clones. Microsatellite markers are being developed which are useful in the analysis of closely-related germplasm, i.e. the types of crosses breeders usually make. A Thai graduate student at Cornell has developed a simple and rapid method for determining the genotype of seeds before germination and without DNA isolation.[17] The analysis is conducted on supernatants from half-seeds preincubated in aqueous solution. The remnant half-seeds having the desired genotypes can then be grown into plants.

The linkage maps of *O. sativa* may also have applications beyond rice. Comparative mapping of cDNA clones with maize and rice, and wheat and rice, demonstrate conservation of linkage relationships across substantial regions of chromosomes in these distantly-related monocot genera.[18] In some instances entire chromosomes or chromosome arms are nearly identical with respect to gene order and gene content. This work suggests that it may be possible to develop a common reservoir of markers and linkage information for several of the important Gramineae crop species, to gain new insights through comparison and to use the rice maps and insert libraries as tools for cloning genes from other cereals.

GENETIC MAPS AND MARKERS OF RICE PATHOGENS

Molecular genetic maps and markers can also be used to analyze and monitor genetic diversity and variability in populations of rice pathogens, thereby facilitating more effective use and deployment of resistance genes. Repetitive DNA elements are especially well-suited for generating informative markers because they allow analysis of a large portion of the genome with one probe.[19]

For the rice blast fungus, *Magnaporthe grisea*, restriction fragment length polymorphisms (RFLPs) generated by DNA probes for dispersed repeated sequences have provided a new fingerprinting mechanism for understanding population dynamics of the fungus and variation in virulence.[20] Based on such DNA fingerprinting, blast pathogen populations can be organized into a discrete number of families with similar genetic backgrounds, referred to as lineages. In the U.S. the rice blast population was shown to be composed of eight well-differentiated genetic lineages with defined pathotype associations.[21]

Using a similar approach, Levy et al[22] conducted fingerprinting analysis for nearly 1,000 field isolates of the blast fungus collected from over a dozen countries throughout the world. They concluded the following: (1) the blast pathogen populations in each country are clonal and composed of a small number (6-18) of lineages; (2) where pathogenicity testing was available, each lineage was associated with one to several closely related pathotypes, or had a limited cultivar range, and these relationships are generalized over several years; (3) rice pathogens and grass pathogens are generally host-limited and easily distinguished by DNA fingerprints; and (4) globally, the majority of lineages appear to be indigenous, but there are at least four lineages that are widespread in the Americas and at least two other lineages are present both in Asia and the Americas.

The relationships between lineage, pathotype and cultivar are now being examined in greater detail at several locations with the aim of developing specific schemes for durable resistance breeding which emphasize lineage-exclusion strategies. Lineage analysis is already being used for evaluating the diversity of pathogen populations in blast screening nurseries. In the Philippines, for example, only five lineages were found at the IRRI blast nursery, while 12 lineages were detected at an upland site elsewhere in the country.[23]

RICE GENETIC ENGINEERING

Transgenic cereal plants were first obtained in rice using protoplast-based transformation systems.[24-27] In these systems uptake of DNA by protoplasts is usually mediated by treatments that increase the permeability of the cell membrane, and regeneration is dependent upon delicate manipulations of both protoplasts and embryogenic cell-suspension cultures. While efficiencies remain low and many of the resulting transgenic plants are pollen-sterile, these techniques have provided the basis for the first field tests of transgenic cereals and for production of numerous transgenic rice plants containing potentially useful alien genes.[28-30]

Biolistic techniques are now also used for rice transformation with promising results, particularly because they appear to be genotype-independent. Christou et al[31] obtained transformants by subjecting 12- to 15-day old rice immature embryos to electric-discharge, particle-mediated transformation with a selectable marker gene. Cao et al[32] used a helium gas particle gun to transform cells in an embryogenic suspension culture with a selectable-marker gene. Transformed calli could be selected and regenerated into plants.

The particle gun-based rice transformation system developed by Li et al[33] at the International Laboratory for Tropical Agriculture (ILTAB) in California was further refined and successfully transferred to other laboratories in the U.S., China and Malaysia through training of key scientists at ILTAB. With Program funding the capacity of ILTAB to provide such training to other developing-country scientists was recently expanded.

Other techniques for the production of transgenic rice have been reported. Dekeyser et al[34] developed a procedure to electroporate DNA into intact and organized rice tissues and reported transient gene expression. This technique has since been refined to give stable transformation of rice and other crops. *Agrobacterium* was shown to transfer its tDNA into rice as measured by Agroinfection.[35] Using a so-called "superbinary" vector Hiei et al[36] recently presented convincing evidence of stable rice transformation mediated by *Agrobacterium*.

Using these various techniques, chimeric gene constructs can now be integrated into the nuclear DNA of rice plants and passed on to subsequent generations as part of the rice genome. The added gene may encode a new protein, or it may alter the level or location of expression, or both, of existing proteins. The site of integration appears to be random. Techniques for targeting genes to a particular site on the rice genome, or for replacing an existing gene with an engineered alternative, are being developed, but are not yet available.

The coding sequence of these chimeric genes can come from any source—rice, wild relatives of rice, other plants, microbes, animals or chemical synthesis. The regulatory sequences will need to function in rice and often will come from rice. Many of the more-sophisticated and powerful uses of genetic engineering will involve highly-regulated genes that are expressed at desired levels in particular cells, tissues or organs at particular stages of development, or in response to particular environmental stimuli. Considerable Program funding has been committed to research and aimed at understanding and utilizing these regulatory mechanisms in rice.[37]

Transgenic plants have been produced with a variety of potentially useful genes (e.g. viral coat-protein genes, herbicide resistance genes, *Bacillus thuringinesis* (Bt) δ-endotoxin genes, male-sterility genes and modified storage-protein genes) and are now being evaluated. Within a few years, transgenic rice plants with useful new genes should provide rice breeders with an expanding source of valuable genetic variability.

RESEARCH ON PRIORITY TRAITS

Substantial Program support is currently committed to research on promising strategies for achieving the desired traits listed in Table 14.4. Both genetic engineering and marker-aided selection strategies are being employed. The international centers and national programs are using their own capacity to work on many other traits. Within the next few years, the first improved rice varieties produced in part through marker-aided selection should reach farmers fields, with many others to follow. Within three years there should be field tests of rice plants containing transgenes for resistance to pests and pathogens. Within five years there should be substantial progress on the abiotic stresses, modification of biosynthetic pathways and hybrid-seed production systems.

INTELLECTUAL PROPERTY RIGHTS AND BIOSAFETY

The advent of biotechnology has caused intellectual property rights (IPR) and biosafety to become more important issues for the international agricultural community than was previously the case. They are important issues for all Rockefeller Foundation science-based development activities.

With regard to IPR, all prospective grantees are informed of the Foundation's IPR policies and are required to submit to the Foundation information on their IPR policies. If there are significant discrepancies, a special agreement is usually negotiated. Rice biotechnology grantees are expected to share materials and technology resulting from Program-supported research with cooperating researchers at zero royalty for use in developing countries. At the same time, grantees are encouraged to pursue intellectual property rights on their discoveries in order to obtain economic return in developed countries for support

Table 14.4. Desired traits for which collaborative research is making good progress and estimates of years to completion

Desired Traits	Years to Completion[a] (less than)
Resistance to tungro virus	3
Resistance to ragged stunt virus	3
Resistance to hoja blanca virus	3
Resistance to yellow mottle virus	3
Resistance to yellow stemborer	3
Resistance to brown planthopper	3
Durable resistance to blast	5
Durable resistance to bacterial blight	5
Durable resistance to sheath blight	5
Drought tolerance	8
Submergence tolerance	8
Cold tolerance	8
Enhanced starch biosynthesis	3
Provitamin A synthesis in grain	8
Hybrid seed production systems	5
Male-sterility genes	3
Restorer genes	3
Markers for predicting heterosis	3

[a]Years to Completion = Estimate of the number of years Program funding will need to continue to provide breeders with a breeding line having the desired trait via biotechnology

of further research and to maintain a strong bargaining position in the event of any intellectual property disputes. To help developing countries formulate their own IPR policies, the Program supports workshops, research and publications. There are IPR presentations and workshops at each annual meeting, and the IPR specialists funded by the Program are available to provide advice.

With regard to biosafety, if the developing countries are to share fully in the benefits of biotechnology, while minimizing risks and guarding against misuse, they must develop and implement biosafety systems which are workable, effective and based on rigorous scientific evaluation. To help them establish such systems professionals from agencies and organizations that have responsibilities and expertise in this area contribute to the Program. Regional workshops on rice biosafety have been supported in Thailand and Indonesia and include input from scientists who have implemented strong biosafety regulations in industrialized countries. Developing-country scientists responsible for biosafety have received training at the USDA Animal and Plant Health Inspection Service and other appropriate locations.

In addition, the Foundation helped to establish the Biotechnology Advisory Commission, a unit of the Stockholm Environment Institute, which serves as an independent resource for impartial biosafety advice. The Commission is designed to help developing countries assess the possible environmental, health and socioeconomic impacts of proposed biotechnology introductions. In addition to general support, the Foundation has agreed to fund specific assessments relevant to rice biotechnology. This assures that the countries and institutions participating in the Program have access to a knowledgeable independent source of advice concerning biosafety issues when and if they need it.

CONCLUSIONS

The International Program on Rice Biotechnology has made good progress toward its objectives. The techniques of biotechnology have been successfully developed for tropical rice. Moreover, because it is well suited for biotechnological manipulations, rice is becoming a model monocot plant for further research. Considerable rice biotechnology research capability now exists in rice-dependent countries and nearly half of the Program's fellows are still receiving training and will soon return home to begin their own research careers. A better understanding now exists of the social and economic consequences of introducing new rice technology and the implications this has for research priorities and policies. New rice varieties resulting in part from anther culture and wide crossing have reached farmers' fields and are contributing to improved rice production. Improved seeds resulting in part from marker-aided selection are not far behind. Rice plants with potentially-useful new traits resulting from genetic engineering are being evaluated and should lead to improved varieties within five years. In the future, rice production should continue to receive a stream of technological innovations, in part resulting from biotechnology, which hopefully will enable food production to keep pace with population growth in Asia until there is a stable population fed by sustainable agricultural systems.

REFERENCES

1. Administrative Committee on Coordination/Subcommittee on Nutrition (ACC/SNC). Second report on the world nutrition situation, Vol. 1, Global and Regional results. Geneva: World Health Organization, 1992.
2. Herdt RW. Research Priorities for Rice Biotechnology. pages 19-54. In: Khush GS, Toenniessen GH, eds. Rice Biotechnology. Wallingford: CAB International, 1991.
3. David CC, Otsuka K Modern Rice Technology and Income Distribution in Asia. London: Lynne Rienner Publishers, 1994.
4. Arumanagathan K, Earle ED Nuclear DNA content of some important plant species. Plant Molec Biol Report 1991; 9:208-218.
5. Causse M, Fulton TM, Cho YG, Ahn SN, Chunwongse J, Wu K, Xiao J, Yu Z, Ronald PC, Harrington SB, Second GA, McCouch SR, Tanksley SD. Saturated molecular map of the rice genome based on an interspecific backcross population. Genetics 1994; in press.
6. Wang G-L, Ronald PC. Construction of a bacterial artificial chromosome (BAC) library for physical mapping and cloning of a bacterial blight resistance gene, Xa-21. In: Abstracts of the Seventh Meeting of the International Program on Rice Biotechnology, New York: The Rockefeller Foundation, 1994:11.
7. Zhao X, Wang G, Zhang HB, Ding X, Woo SS, Zhang Z, Yang J, Patterson AH, Wing RA. Megabase DNA isolation and BAC/YAC Cloning in Rice (*Oryza sativa* L). In: Abstracts of the Seventh Meeting of the International Program on Rice Biotechnology, New York: The Rockefeller Foundation, 1994:16.
8. Wang ZY, Second G, Tanksley SD. Polymorphism and phylogenetic relationships among species in the genus *Oryza* as determined by analysis of nuclear RFLPs. Theor Appl Genet 1992; 83:565-581
9. Lampe KJ. Report of the Director General, International Rice Research Institute. Manila: IRRI 1993; 3:(1):8-11.
10. McCouch SR. Progress in RFLP gene tagging and marker-aided selection. A paper presented at the Asia-Pacific Conference on Agricultural Biotechnology, Beijing, China, August 20-24, 1992.
11. Ahn SN. Comparative mapping of rice and maize genomes and mapping quality genes of rice using restriction fragment length polymorphism (RFLP) markers. Ph.D. Dissertation. Ithaca: Cornell University, 1993.
12. Ahn SN, Bollich CN, McClung AM, Tanksley SD. RFLP analysis of genomic regions associated with cooked-kernel elongation in rice. Theor Appl Genet 1993; 87:27-32.
13. Ludlow MM, Lilley JM. Screening for genetic variation in osmotic adjustment, epidermal conductance and dehydration tolerance in rice. In:

Abstracts of the Sixth Annual Meeting of the International Program on Rice Biotechnology. New York: The Rockefeller Foundation, 1993:145.

14. Yu LX, Ray JD, Wang G, Champoux M, McCouch SR, Nguyen HT. Identification of molecular markers linked to root penetration in rice. In: Abstracts of the Sixth Annual Meeting of the International Program on Rice Biotechnology. New York: The Rockefeller Foundation, 1993:146.

15. Champoux MC, Wang GL, Sarkarung S, Mackill DJ, O'Toole JC, Huang N, McCouch SR. Locating genes associated with root morphology and drought avoidance via linkage to molecular markers. Theor Appl Genet 1994; in press

16. Zhang Q, Gao YJ, Yang SH, Ragab RA, Maroof MAS, Li JX, Li ZB. Molecular marker-based analysis of heterosis in hybrid rice. In: Abstracts of the Seventh Meeting of the International Program on Rice Biotechnology, New York: The Rockefeller Foundation, 1994:5.

17. Chunwongse J, Martin GB, Tanksley SD. Pre-germination genotypic screening using PCR amplification of half-seeds. Theor Appl Genet 1993; 86:694-698.

18. Ahn S, Anderson JA, Sorrells ME, Tanksley SD. Homologous relationships of rice, wheat and maize chromosomes. Mol Gen Genet 1993; 241:483-490.

19. Harner J, Farrall L, Orbach M, Valent B, Chumley F. Host species-specific conservation of a family of repeated DNA sequences in the genome of a fungal plant pathogen. Proc Natl Acad Sci USA 1989; 86:9981-9985.

20. Hamer JE. Molecular probes for rice blast disease. Science 1991; 252:632-633.

21. Levy M, Romao J, Marchetti MA, Hamer JE. DNA fingerprinting with dispersed repeated sequence resolves pathotype diversity in the rice blast fungus. Plant Cell 1991; 3:95-102.

22. Levy M, Correa-Victoria FJ, Zeigler RS, Shen Y, Shajahan AKM, Gnanamanickam SS, Nelson RE, Manry J, and Hamer JE. International Atlas of Genetic Diversity in the Rice Blast Fungus 1992. In: Abstracts of the Sixth Annual Meeting of the International Program on Rice Biotechnology. New York: The Rockefeller Foundation, 1993; 110.

23. Leung H, Shi Z, Nelson R, Bonman M, Estrada B, Chen D, Bernardo M, Scott R, Zeigler R. Analysis of populations of the rice blast fungus at upland rice screening sites. In: Abstracts of the Sixth Annual Meeting of the International Program on Rice Biotechnology. New York: The Rockefeller Foundation 1993; 184.

24. Toriyama K, Arimoto Y, Uchimiya H, Hinata K. Transgenic rice plants after direct gene transfer into protoplasts. Bio/Technology 1988; 6:1072-1074.

25. Zhang W, Wu R. Efficient regeneration of transgenic plants from rice protoplasts and correctly regulated expression of the foreign gene in the plants. Theor Appl Genet 1988; 76:835-840.

26. Yang H, Zhang HM, Davey MR, Mulligan BJ, Cocking EC. Production of kanomycin resistant rice tissues following DNA uptake into protoplasts. Plant Cell Rept 1988; 7:421-425.

27. Peng J, Kononowicz H, Hodges TK. Transgenic indica rice plants. Theor Appl Genet 1992; 83:855-863.

28. Datta SK, Datta K, Soltanifar N, Donn G, Potrykus I. Herbicide-resistant indica rice plants from IRRI breeding IR72 after PEG-mediated transformation of protoplasts. Plant Mol Biol 1992; 20:619-629.

29. Hayakawa T, Zhu Y, Itoh K, Kimura Y, Izawa T, Shimamoto K, Toriyama S. Genetically engineered rice resistant to rice stripe virus, an insect-transmitted virus. Proc Natl Acad Sci USA 1992; 89:9865-9869.

30. Rathore KS, Chowdhury VK, Hodges TK. Use of *bar* as a selectable marker gene and for the production of herbicide-resistant rice plants from protoplasts. Plant Molec Biol 1993; 21:871-884.

31. Christou P, Ford TL, Kofrom M. Production of transgenic rice (*Oryza sativa* L) plants from agronomically important indica and japonica varieties via electric discharge particle acceleration of exogenous DNA into immature zygotic embryos. Bio/Technology 1991; 9:957-962.

32. Cao J, Duan X, McElroy D, Wu R. Regeneration of herbicide resistant transgenic rice plants following microprojectile-mediated transformation of suspension culture cells. Plant Cell Rept 1992; 11:586-591.

33. Li L, Qu R, de Kochko A, Fauquet C, Beachy RN. An improved rice transformation system using the biolistic method. Plant Cell Report 1993; 12:250-255.

34. Dekeyser RA, Claes B, DeRycke RMV, Habets ME, Van Montagu MC, Caplan AB. Transient gene expression in intact and organized rice tissues. Plant Cell 1990; 2:591-602.

35. Raineri DM, Bottino P, Gordon MP, Nester EW. Agrobacterium-mediated transformation of rice (*Oryza sativa* L.). Bio/Technology 1990; 8:33-38.

36. Hiei Y, Ohta S, Komari T, Kumashiro T. Efficient transformation of rice (*Oryza sativa* L.) mediated by Agrobacterium and sequence analysis of the boundaries of the T-DNA. Plant Journal 1994; 6:271-282.

37. Izawa T, Foster R, Nakajima M, Shimamoto K, and Chua NH. The rice bZIP transcriptional activator RITA-1 is highly expressed during seed development. Plant Cell 1994; 6:1277-1287.

CHAPTER 15

PLANT BIOTECHNOLOGY FOR SMALL-SCALE AGRICULTURE

Bert Visser and Hans Wessels

The views expressed in this text are those of the authors and do not necessarily reflect DGIS policies.

INTRODUCTION AND OBJECTIVES

In 1992 the Netherlands Minister for Development Cooperation established the Special Programme Biotechnology and Development Cooperation. Like similar efforts, the Programme aims to increase developing countries' access to biotechnological expertise and innovations, and to contribute to the solution of developmental problems. To achieve this, support for relevant international coordinating activities and networks as well as the integration of the development dimension in biotechnology policies in the Netherlands form part of the Programme's agenda. But most of all, technical cooperation is a major activity of the Programme. In developing technical cooperation, the Programme exhibits some additional features that are rather specific for this particular effort.[1]

First, the Programme focuses on small-scale agriculture in an attempt to contribute directly to poverty alleviation. In doing so, it tries to identify activities which are not taken up by the international research community. Second, the Programme is not a research program per se, but features both research and development components. It is not interested in laboratory products ending on the shelves, but in products used in the field. Third, the development of program activities is seen as an initial responsibility of the developing country. Although the specific approach is an integral and essential part of the Programme, no bias concerning the choice of projects exists. Fourth, to assure farmers' use of biotechnology products, it regards farmers participation in problem identification, priority setting, project formulation and project implementation as essential. Fifth, capacity building is a necessary, albeit not an autonomous, complementary objective of the Programme in an effort for closing the still-widening gap between the North and South in the development and use of biotechnology. To contribute to capacity building and to an enabling environment for the application of biotechnology, topics like intellectual property rights and biosafety and tools like international research networks receive specific attention.

Finally, following the objective to work from farmers' experiences and to contribute to local capacity building, technical cooperation is being developed with four countries, i.e. Zimbabwe, Kenya, Colombia and India. In those countries specific regions are designated target areas to organize end-users' involvement in program development. This approach should enable the Programme to realize farmers' participation, but is not intended to eliminate a subset of project options.

The Special Programme has been set up for an initial period of 5 years. Outside experts have been brought in for the development of the Programme in the priority areas. An Advisory Board consisting of individuals representing relevant specializations, experience and social sectors was installed. For the implementation of the Special Programme an amount of Dfl. 50 million is available over 5 years, excluding staff and office expenditures in the Hague.

A PARTICIPATORY APPROACH: INSTITUTIONAL ASPECTS

Characteristic for the Special Programme is that activities are identified via a participatory process that starts with the end-users and beneficiaries of the relevant technology. It means that projects are formulated on the basis of a local needs' assessment and priority-setting process, in which end-users, researchers and policy-makers are involved. This approach has been coined the "interactive bottom-up approach."[2] The rationale behind the approach is that an understanding of farmers' problems, their interests and farming systems is crucial in the assessment of the appropriateness of biotechnological innovations. A central element of the approach is that information on the appropriateness, as well as on the feasibility and comparative advantage of innovations, is assimilated, discussed and integrated in the framework of the Programme and in its collaborative projects.

To illustrate how the Programme attempts to realize this approach, a summary of the activities preparing the collaboration with Kenya is given. A first activity was to obtain an overview of the agricultural sector, agricultural-research efforts, available human resources, institutional capacity and national policies in the area. A country study, prepared by Dr. Olembo, served this goal.[3] Extensive discussions of the Programme staff with representatives of government, research institutions, universities, NGOs and farmers' organizations further substantiated the data of the country study. As a second step a local consultant, ETC Kenya, was hired to act as a local representative of the Programme, and to assist in the following phase of program development. This phase involved the consultation of small-scale farmers in two districts, representing two dominant agro-ecological zones in Kenya, i.e. semi-arid Machakos and high-producing but densely-populated Kakamega. Farmers' workshops were convened and addressed the farming systems and priority crops in the two districts, the problems that farmers encountered in crop cultivation and the selection of those problems that might be amenable to a biotechnological approach. Also, the role and functioning of the various institutions (government, companies, NGOs, farmers' organizations, etc.) relevant to agriculture in the districts were analyzed.

In parallel and reacting to the findings of the farmers' workshops, workshops of researchers and policy-makers were held. Researchers convened to discuss the potential contribution of biotechnology to small-scale agriculture given available or readily-achievable Kenyan resources (infrastructure and human resources) and the international state-of-affairs in biotechnology. Separate workshops were dedicated to opportunities in crops, livestock and agroforestry. Policy makers convened to examine how the Programme activities might fit into national policies and how it might stimulate a further development of policies.

Subsequently a national workshop was organized in which participants from the farmer, research and policy workshops came together, in order to assimilate and compare the information and conclusions from the previous workshops. Nine farmers joined in a meeting of 45 participants in total. The national workshop comprised plenary presentations and discussions as well as small-group discussions in parallel sessions. On the last day, the outcome and conclusions of the four-day workshop were presented to a larger audience of government officials, researchers of national and international institutes and private enterprise.[4]

As a result of the national workshop, the Kenya Agricultural Biotechnology Platform was established. This Platform consists of ten members who participate in their personal capacities, although they were chosen to cover the various institutions such as ministries, research institutions, universities, NGOs, farmers' organizations and farmers' communities. The Platform focuses on problem-oriented application of biotechnology for small-scale producers. It attempts to link biotechnology research and policy, by identifying priority areas for agricultural biotechnology in more detail, and by identifying executing institutions and organizations in Kenya. It advises the Special Programme on the relevance of the activities generated within the Programme's framework and will monitor project progress. In addition, the Platform will be involved in the promotion of agricultural biotechnology in Kenya through workshops, conferences and the publication of a newsletter, and it will be engaged in the organization of "feed-back" consultations with the prospective end-users. The Platform published a brochure containing a call for proposals based on the broad priority areas defined in the national workshop.[5]

The first meetings between Programme staff and key persons in this process took place in October 1992, the national workshop was held in September 1993, and the brochure appeared in July 1994. Proposals are now submitted and reviewed. This brings the Programme from the phase of problem identification and priority setting into the next phase of project formulation. Some expectations concerning the next phase are given further in this chapter.

In Zimbabwe and Colombia, similar processes have been initiated. Steering committees, largely with the same scope and constitution, have been established. In Zimbabwe, a national workshop was organized in May 1993,[6] and in Colombia such a conference has been planned for early 1995. In Colombia much attention has been paid to farmers' participation in the selected target region, the Atlantic Coast. A permanent committee of small-scale farmers in the region has been installed. A detailed socio-economic study is under way to provide a solid basis for discussions on appropriateness and feasibility of project options. In Zimbabwe, farmers' participation has been weak, but is now strengthened both at the level of the steering committee and through additional socio-economic studies in the targeted regions, Hwedza and Buhera districts encompassing an array of agro-ecological zones. In the state of Andra Pradesh in India, farmers participation is sought through collaboration with two active NGOs, i.e. Youth for Action in the district Mahboobnagar, and the Sri Aurobindo Institute of Rural Development in Nalgonda. Multidisciplinary assessments focusing on the farming systems will contribute to priority setting.

In all four countries informal groups will play an important role in detailing Programme priorities and formulating project proposals. These groups may consist of researchers from national institutions as well as international counterparts, and of farmers and their representatives. Strategies will have to be developed to cope with complex institutional project structures due to the objective to cover both research and development, and to involve both researchers and users of technology. Detailed project ideas will be discussed with farmer groups involved in the earlier stages of Programme development to obtain necessary feedback, e.g. on the choice of projects based on the identified priority areas, on the choice of technologies and of crop varieties and properties that will form the object of project proposals or on the distribution and adaptation aspects of improved agricultural inputs.

APPROPRIATE BIOTECHNOLOGIES

Within the context of this book it may not be necessary to discuss in detail the type of biotechnologies which may contribute to improvements in small-scale agriculture. However, it seems useful to allude briefly to differences in the interpretation of the concept of biotechnology, which has its implications for the scope of the Programme. In addition, some preferences based on the relevance of different technologies in the research infrastructure and agriculture of developing countries are discussed.

The term biotechnology has been coined in the scientific world of the seventies to describe novel developments in biological research and its applications, no longer only making use of intact living organisms but also of parts thereof. Many researchers generally will use this interpretation of biotechnology, to include mainly modern developments at the cellular and molecular level. Plant cell and tissue culture has become widely used in recent decades all over the world.[7] More recent developments in biotechnology are based on molecular methods, and concern in particular the isolation, characterization, modification and transfer of genes, as well as the targeted use of specific DNA sequences as markers in plant breeding. Applications of these methods are only beginning to emerge, and need to be integrated with organismal and cellular techniques.

Alternatively, biotechnology can be regarded to encompass all technologies based on biology either at the level of the organism, the cell or the molecule. Under this broad interpretation many applications including animal and plant breeding as well as rural practices in food processing, production of biofertilizers and biopesticides, ethnoveterinary practices and strategies for the management of genetic resources can be viewed as applications of biotechnology at the level of the organism. These applications have been used for centuries and have often developed outside the laboratory. This second interpretation of the term biotechnology can be encountered in developing countries and in development cooperation.

The Special Programme does not hold a restrictive view on biotechnology. Although it has been established to enable access to modern biotechnology, it realizes that several rural practices involving "classical" biotechnology might be improved using simple means with the help of knowledge and tools stemming from modern biotechnology. Biofertilizers and biopesticides may form candidates for improvement in this category.

In situations of severely-restrained resources as in many developing countries, major benefits can be realized from added investments in the more established biotechnological approaches at the organismal and cellular level. Approaches making use of the molecular techniques will need more time and investments before results will appear. In the discussions about available technologies and desired priorities, applications of molecular biology is therefore sometimes de-emphasized, largely because of the long "incubation time" leading to application. Nevertheless, various potential applications in the molecular area may warrant investments because these aim to cope with problems not effectively addressed by other means. Also, recent years have shown sudden breakthroughs and quick progress in several technical areas, making realistic today what was considered of long incubation time yesterday.

In this context, mass propagation and disease elimination using tissue culture, anther culture for breeding purposes and the selection and improvement of micro-organisms for use as biofertilizers or biopesticides are often regarded as relatively easy, straightforward and short-term project options. In this view the generation of transgenics and the use of molecular markers to improve breeding represent longer-term options. Although this view may be correct as far as the research components of a project are concerned, the picture is far more complicated if one attempts to address the continuum of research and (product) development. Mass propagation and disease elimination through tissue culture is an attractive and effective research strategy mostly for vegetatively-propagated crops, like root and tuber crops and bananas, or crops with a long generation time like fruit trees and agroforestry species. However, because for the Special Programme the aim of a project is not only to establish mass-propagation facilities, but also the distribution and use of the generated plant material, such a project might become more complicated. For most of the above-mentioned crops no effective markets and distribution channels for planting material may exist, meaning that these have to be established in the framework of the project. On the other hand improvement of cereals and legumes using marker-assisted breeding may be an enormous challenge and a lengthy task from a scientific point of view, but the products of such efforts may simply fit into existing marketing and distribution channels, replacing existing closely-related varieties.

In other words, some options that need relatively little and short-term research, may need long lead-times in the adoption phase, whereas options that need comparatively high research investments, may experience easy and quick adoption. Also, it must be noted that these options may be related. Experience with tissue-culture work for mass propagation will greatly benefit and accelerate later projects for the production of transgenics. An additional aspect is the notion that the second type of options involving longer and higher research investments may contribute far more to capacity building in the country involved, provided that a sufficient research base in science and agricultural research is in place to allow for an efficient development of biotechnology. In our view, both short-term and long-term categories need attention, and the problems in agriculture should direct our choices.

PRIORITIES AND UPCOMING TECHNICAL PROJECTS

Most of the efforts of the Special Programme have been invested in the approach described above, and only recently activities under the Programme have begun to shift from farming-systems analysis, problem identification and priority setting towards project formulation. Therefore, information about future project activities is necessarily still limited.

In Zimbabwe, five clusters of crops were identified as priorities, encompassing large sectors of total agricultural production. These clusters are maize, small grains, legumes, vegetables and root and tuber crops. Post-harvest losses were indicated as a vital area for project activities.[6] In a preliminary phase of the Programme two projects were started at the University of Zimbabwe. One involves a cassava regeneration and transformation project aiming at coat protein-mediated resistance against cassava mosaic virus. The training component of this project involved cooperation with research groups in the Netherlands and the U.S. A second project involves a M.Sc. course in biotechnology, which is co-sponsored by SAREC, focuses on plant and microbial biotechnology and is organized in collaboration with Dutch and Swedish universities. A small project studying the genetic variation in sorghum landraces and in ex situ collections in Zimbabwe through molecular markers has been recently carried out. The options for a mass-propagation project in Irish potato and sweet potato are discussed, among others. Finally, the Special Programme brought together Zimbabwean researchers and researchers of KARI in Kenya and of CIMMYT, to assess the feasibility of a project aiming at marker-assisted improvement of local Zimbabwean and Kenyan maize varieties for drought tolerance and insect resistance, and to establish a regional facility to apply the marker technology for breeding purposes.

In Kenya, the national workshop confirmed the development of pest and disease resistant and drought-tolerant varieties of cereals, tuber crops and pulses as priorities. The efforts to study the improvement of local maize varieties were based on this recommendation. In addition, cassava, sweet potatoes, Irish potatoes, fruit trees and agroforestry species have been identified in the priority-setting process as candidates for mass propagation. As a first option, the feasibility of mass propagation of banana presently is discussed to cope with the shortage of disease-free high-quality planting material. In particular, the growing threat of black Sigatoka on top of the damage by nematodes warrants a fast introduction of improved varieties from various sources. The improved use of biofertilizers and the improvement of post-harvest technology also were mentioned as priorities. An optimized, scientifically-sound selection of *Rhizobium* strains effectively nodulating and fixing nitrogen for a broad range of cultivated *Phaseolus vulgaris* varieties, the dominant Kenyan legume, may be a third attractive option, if this can be linked to an efficient and cheap distribution of the nodulants. Much is dependent on the response to the call for proposals distributed by the Kenyan Platform.

In Colombia, the priority-setting process started in 1994, but given the characteristics of the dominant farming systems in the Atlantic Coast high priority can be expected for plantain. Options for a mass-propagation project in plantain, along similar lines as for banana in Kenya, presently are discussed. One question to be answered is whether farmers will be willing and able to buy the improved disease-free material, once the mass-propagation activities will have been fully established and will lack further outside support. The answer to this question determines the long-term sustainability of such activities.

In India, the option of agricultural wasteland rehabilitation through agroforestry and the planting of indigenous fruit trees, to be mass-propagated in tissue culture, has been considered in the initial discussions on priority setting. Identification of other options needs to await further program development.

In summary, considerable efforts will probably be devoted to the area of mass propagation and disease elimination through tissue culture. Both agricultural problems and the need to contribute to basic capacity building warrant that expectation. However, given the outcome of the priority-setting processes it may also be expected that proposals using alternative biotechnological tools, like marker-assisted breeding and improvement of microbial agents, and maybe also genetic modification of crops, will be forwarded.

POLICIES AFFECTING ACCESS TO BIOTECHNOLOGY

The probability of success of biotechnology projects is not only dependent on the appropriateness of the offered improvement or on the scientific, technical and economic feasibility of project objectives. The probability of success is also dependent on the socio-political environment in which projects take place and in which the use of project products is propagated. Specific topics concerning access to and benefit of biotechnology include safety issues, public acceptance and property rights on improved materials. Dealing with the issues of biosafety and property rights is seen as a prerequisite for technical cooperation involving transgenics.

Policies on biosafety must address the establishment and the implementation of biosafety mechanisms. Whereas many developed countries have adopted biosafety mechanisms, only few developing countries have yet followed suit. However, currently recombinant DNA organisms and transgenic crops are introduced or developed in a growing number of developing countries, both for contained use and for deliberate release into the environment. This necessitates a quick response by developing countries, which can benefit from the experience that has been obtained in the North. Rather than simply copying existing mechanisms, developing countries should establish their own biosafety mechanisms. These should be adjusted to their legal and political systems, the infrastructural and human capacity available to implement such mechanisms, the agro-ecological specifics and the magnitude of the demand in these countries. In building their biosafety mechanisms, countries might strongly benefit from regional cooperation. From this perspective the Special Programme has been involved, in cooperation with the Research Council of Zimbabwe, in the organization of the African Regional Conference for International Cooperation on Safety in Biotechnology, which was held in Harare in 1993,[8] and in cooperation with IICA, CIAT and the U.S. agencies, in the organization of the Workshop on the Harmonization of Biosafety in the Americas, which aimed at capacity building in the Andean region and was held in Cartagena in 1994. The Special Programme also supports a regional focal point for Southern and Eastern Africa, to be established in Harare. This focal point shall serve sharing of expertise and of data concerning risk assessments and monitoring in the region and which can function as a regional node of communication to global information systems.

Intellectual property rights will have a big impact on access to products of modern biotechnology. National attitudes towards these rights may affect international trade, and in the longer term also affect the potential domestic use of protected products in technology transfer and foreign investments. Many farmers and NGOs in developing countries are opposing intellectual property rights. In that context, provisions in plant breeders' rights allowing the privilege to farmers to keep their own seed stocks may be of much relevance to the small-scale agricultural sector. The Special Programme has fostered discussions about the impact of plant breeders rights and patent rights on national agriculture in developing countries and on the small-scale sector in particular. A recent study supported by the Programme of the effects of plant breeders rights in Latin America indicated that plant breeders rights benefit national breeders rather than the multinational breeding companies.[9] The rationale for this notion is that multinational companies almost exclusively produce hybrid varieties which are "protected" by biological means, whereas national breeders tend to produce open-pollinating varieties.

THE ROLE OF NETWORKS

Because of limited resources the national research base in biotechnology in many developing countries is small, whereas access to global expertise and information is often poor. In the view of the Special Programme networks and targeted information services may help developing countries to increase progress in research and to improve its quality. For this reason the Programme promotes and sponsors such activities.

From 1988 through 1992 The Netherlands chaired the CGIAR (Consultative Group on International Agricultural Research) Task Force Biotechnology (BIOTASK). This Task Force was created to further the integration of biotechnology in the CGIAR with special attention for the socio-economic aspects of the introduction of biotechnology. The Task Force consisted of representatives of donor agencies, research institutes and developing countries.

The Special Programme as the chair of an international donor group has been responsible for the foundation of the Intermediary Biotechnology Service (IBS) at ISNAR in The Hague. The IBS is a demand-driven, problem-oriented neutral advisory service with the objective to advise developing countries on policy aspects of biotechnology and on the integration of biotechnology in their research programs, and to assist in identifying major problems amenable to biotechnological approaches. Furthermore, the IBS aims at identifying and mobilizing biotechnological expertise for national programs in developing countries.

Based on the recommendations of the workshop Cassava and Biotechnology, which was held in Amsterdam in 1990, the Cassava Biotechnology Network (CBN) was set up at CIAT. Cassava is the fourth largest food crop in the world, providing food for 500 million people. The CBN aims to function as a forum for cassava biotechnology and aims at stimulating cassava biotechnology research and at integrating the priorities of small-scale cassava producers in the biotechnology research agenda.

An important activity of the Special Programme has been the publication since 1989 of the quarterly journal *Biotechnology and Development Monitor*, a joint activity with the University of Amsterdam. The *Monitor* is directed at researchers and policy makers in developing countries as well as in developed countries and has a circulation of around 5,000 copies distributed free-of-charge to subscribers. Right from the start of the project high priority was given to an increasing participation from developing countries and to a final transfer of the project to either an organization in a developing country or an international organization. Regular contributions are obtained from ACTS in Kenya, RIS in India and IICA in Costa Rica.

Complementary to the mostly policy-oriented overviews of the *Monitor*, an information service on recent scientific developments in the field of biotechnology has been made available. In line with the recommendation of BIOTASK, the Special Programme has procured 150 free subscriptions of *AgBiotech News and Information* for the libraries of research institutions in developing countries for a total period of three years. This publication from the Commonwealth Agricultural Bureau International (CABI) can be regarded as a highly-informative technical-abstract journal.

Finally, the Programme supports CAMBIA, the Center for Application of Molecular Biology in International Agriculture, based in Canberra. Its services will be targeted at the needs and constraints of scientists working in developing countries and at developing and providing appropriate tools to solution of local problems.

STRENGTHS AND WEAKNESSES OF THE APPROACH

At this stage only few preliminary remarks can be made on the strengths and weaknesses of the strategy to develop a demand-driven, product-oriented biotechnology program for the small-scale agricultural sector. The end-users' involvement, the wealth of data this involvement provided about the farming systems, the true dialogue between end-users, researchers and policy makers, a resulting change in perspective and a commitment of all participants involved in program development can all be regarded as the strengths of the approach. As a consequence of the approach, a large amount of time required to organize the involvement and commitment of all participants should be anticipated. Weaknesses concern the difficulties in focusing from general priorities down to project ideas and the impossibility to incorporate the dynamics of agricultural needs and scientific developments into the prioritization process.

Farmer workshops in Kenya and Colombia enabled internal consultations between farmers, independent from research agendas. The farmers concluded which problems they experienced as the major ones. These conclusions in turn formed a guideline and checklist for the researchers preparing contributions to the priority-setting process. The followed procedure resulted in an active participation of the farmers in the National Workshop in Kenya, including statements on the results of the farmer workshops. Representatives of the various sectors communicated directly, and in doing so, learned to comprehend each other's views on reality and to value each other's expertise. A change in perspective amongst all participants on prioritizing research followed. It became increasingly clear that end-users have an important role to play in developing research agendas. Both in Kenya and Zimbabwe, where the process has sufficiently advanced, prioritizing was based in the first place on the problems formulated by the farmers, and not on the basis of technological options and opportunities. This process led to the collective identification of priority areas. It warrants the expectation of a stronger commitment during the further development of the cooperative programs.

The approach has produced an enormous amount of data and detailed insight in the perspectives of the various groups involved, which might benefit other cooperative programs in agriculture and biotechnology. The process led to the understanding that indigenous knowledge and practice is important in the planning of the research agenda and the development of research products. The designation of target areas for program implementation was instrumental in obtaining farmers' participation. At the same time, one should realize that focusing on these target areas will have to be reconciliated with due attention to the major problems of the country's small-scale agricultural sector where obvious opportunities exist.

On several occasions it appeared necessary to explain the expected added value of the approach followed: a program aimed at problem solving through appropriate solutions, to which end-users are committed. Questions specifically concerned the time frame. This approach takes much time and effort in comparison with conventional approaches. A long lead-time may result in loss of commitment, and the donor may appear hesitant. Therefore, the continued involvement of all major players in the process requires constant attention, and an on-going supply of information on the state-of-affairs in program development is essential. As a result of the heterogenous background of participants in the priority setting, it appeared unrealistic to complete the priority setting down to the level of individual project ideas at the workshops. The established steering committees will have the task to identify project proposals that best fit the priority areas identified during the workshops. Also, the results of the general priority setting were predictable and strongly based on an analysis of the current problems and conditions. It appeared difficult to take into account future trends in agriculture and demography, and in technological progress.

THE CHALLENGE OF THE PROGRAMME

It remains to be decided whether a close and motivated involvement of the various sectors in the next phase, project formulation, will lead to projects of higher quality, in which obvious mistakes have been avoided. The question whether the outcome of the process justifies all efforts can only be answered by a comparison of the quality of the cooperative program based on this approach with other programs in this area. For the Special Programme the central criteria will be whether results really benefited the end-users in the small-scale sector and whether products have really reached the fields.

REFERENCES

1. Biotechnology and development cooperation. Priorities and Organisation of the Special Programme. The Hague, the Netherlands: Directorate General International Cooperation, Ministry of Foreign Affairs, 1992.
2. Bunders JFG, Broerse JEW. Appropriate biotechnology in small-scale agriculture: how to reorient research and development. Oxon, UK: CAB International, 1991.
3. Olembo NK. Agricultural biotechnology Kenya: a country study. The Hague, the Netherlands.: ISNAR, 1991.
4. Wekundah JM, Visser B. Proceedings of the National Agricultural Biotechnology Workshop. Nairobi: Kenya Agricultural Biotechnology Platform, 1994.
5. The Kenya Agricultural Biotechnology Platform Brochure. Nairobi: ETC Kenya, 1994.
6. Chikwamba R, Feresu SB. Proceedings of the workshop: prioritizing the biotechnology agenda for Zimbabwe. Harare: ENDA, 1993.
7. Hardy RWF. Overview of opportunities in biotechnology. In: ATAS Issue 9 Biotechnology and Development. New York: United Nations, 1992.
8. Van der Meer PJ, Schenkelaars P, Visser B, Zwangobani E. Proceedings of the African Regional Conference for International Cooperation on Safety in Biotechnology. Directorate General International Cooperation. The Hague, the Netherlands: Ministry of Foreign Affairs, 1993.
9. Van Wijk J. Intellectual property protection with regard to plant genetic engineering in developing countries. Abstracts 4th International Congress of Plant Molecular Biology, no. 2113. Amsterdam: 1994.

THE ROLE OF JIRCAS IN INTERNATIONAL TECHNOLOGY TRANSFER RELATED TO BIOTECHNOLOGY APPLICATION TO AGRICULTURE AND FOOD PROCESSING IN JAPAN

Keiji Kainuma

PHILOSOPHY AND ROLE OF JIRCAS[1]

Since its establishment in 1970, the Tropical Agriculture Research Center (TARC) has played a major role in the development of agriculture and forestry in the tropics and subtropics, which is of interest to many lesser-developed nations. The achievements stemming from collaborative research carried out with a large number of research institutions in various countries generally have been highly evaluated. The reorganization of TARC into a new center, Japan International Research Center for Agricultural Sciences (JIRCAS), is an important aspect of Japan's contribution to the welfare of the international community through the promotion of research for technology development in the field of agriculture. The research activities of the new Center will encompass all fields of agriculture, including fisheries for the first time, while forestry will be further promoted. This expansion of the research fields could not have been achieved without the strong participation of all institutions affiliated with the Ministry of Agriculture, Forestry and Fisheries (MAFF). Future research activities of the new Center will not be limited to collaboration with the developing countries in the tropics and subtropics, but will cover wider geographic zones such as cool regions and the temperate zone to deal with problems faced by countries located in the Andean highlands of South America, Central Asia and Mongolia and the northwestern part of China. As a result of the expansion of research fields and targeted regions, the concept of research will be substantially modified along the following lines.

"FOCAL" APPROACH TO RESEARCH VERSUS "COMPREHENSIVE" APPROACH

TARC considered the problems related to agriculture on the basis of individual disciplines or by combining related fields of research. However, to address problems on a global scale, such as sustainable production of food, environmental issues, drought, desertification of arable land or grassland due to overgrazing and the destruction of tropical forests, it is important to promote collaborative research and to exchange data through the systematization of technology by organizing multi- and inter-disciplinary teams involving several research divisions of the Center. For example, studies on the progression of desertification due to overgrazing may require the participation of a soil scientist, geologist, grassland botanist and weed scientist as well as a remote-sensing specialist.

For expansion of the research fields, the planning for projects will obviously be modified. At the beginning, socio-economic studies will be carried out to identify the research priorities of the counterpart countries. In addition, for the systematization of the technology to be transferred, all the research components will be closely linked.

DEVELOPMENT OF A RESEARCH STRUCTURE AT THE CENTER IN JAPAN

To support the research activities carried out overseas, it is essential to improve the structure of research at the Center in Japan. For example a system will be set up to enable the researchers to carry out basic advanced studies related to biotechnology fields or to construct models to simulate various biological processes.

DEVELOPMENT OF INFORMATION SYSTEMS

It is important to develop information systems in order to carry out research collaboration with the respective countries more effectively by compiling and analyzing data relating to agriculture, forestry and fisheries in the developing regions. Through the harmonization of information obtained from socio-economic studies with data stemming from research in fields related to natural sciences, it should become possible to conduct research that would be more comprehensive and cover a wider range of disciplines.

PROMOTION OF CLOSER COLLABORATION WITH VARIOUS RESEARCH ORGANIZATIONS

For the effective implementation of research, it is essential to promote a close collaboration at various levels, including national organizations of the respective counterpart countries, international centers affiliated with the CGIAR, research organizations from developing countries and, in Japan, the administrative authorities and related institutions of MAFF and other research organizations as well as the Japan International Cooperation Agency (JICA).

IMPLEMENTATION OF VISITING RESEARCH FELLOWSHIP PROGRAM AND FURTHER PROMOTION OF TRAINING PROGRAMS

The implementation of the visiting research fellowship program at the Okinawa Subtropical Station of the Center and other training activities reflected the importance of developing human capability within the framework of the contribution of Japan to the welfare of the international community.

It should be emphasized that through the implementation of research activities related to agriculture, forestry and fisheries, the Center is fully committed to contributing to sustainable food production compatible with the preservation of the environment on a global scale so as to upgrade the living conditions of the people in the world. Toward this end, JIRCAS can serve as a means to introduce Japanese advances in biotechnology in agriculture and food processing through training programs and collaborative international research.

PROGRESS OF BIOTECHNOLOGY FOR AGRICULTURE AND FOOD PROCESSING IN JAPAN AND ITS POTENTIAL IN TECHNOLOGY TRANSFER TO DEVELOPING COUNTRIES

Over 15 years have already passed since biotechnology made a brilliant debut in the field of advanced technological research. Its commercialization proceeded at a pace faster than expected at the time when basic research was just initiated. Along with the trend to commercialization, biotechnology at present is progressing steadily and continuously compared to the situation in the initial period full of dreams and fantasies. Biotechnology in the area of agriculture draws a great deal of expectation as an innovative technology in many fields such as food production, livestock, fisheries and food processing.

Against this background, biotechnology is finally entering the phase of industrialization in the advanced countries, with attention being paid to adaptation of biological products developed by biotechnology to the environment or impact on the natural ecosystems. The current situation of biotechnology in the fields of plant and food sciences is described as follows.[2,3]

PLANT BIOTECHNOLOGY

Cell and tissue culture is the basis for plant biotechnology, and often has been used commercially. The most active cell division takes place in the plant apical meristem, where cells are not infected with viruses. To use this characteristic effectively, the apical meristem is cultured aseptically to produce virus-free strawberry, potato and sweet potato plants. This method is referred to as apical-meristem culture. It is interesting to note that strawberries of extremely high quality sold in Japan have been produced by this method. This technique is used to mass-produce western orchids such as *Symbidium* and *Cattleya*, babies' breath (*Gypsophila*) and carnation. Thousands of individuals are produced from one shoot apex by micropropagation.

Embryo-culture methods for growing plant embryos in vitro has enabled the production of hybrids among remotely related plants, which could not have been obtained by conventional fertilization. Remarkable achievements have been recorded, for example, in the production of the interspecific hybrid "Hakuran" between Chinese cabbage and cabbage and the interspecific hybrid "Senposai" between cabbage and *Brassica chinensis* komatsuna, as well as with lily, strawberry and citrus fruit in recent years.

Plants with vegetative propagation can be bred without fixed genetic manipulation, if they are superior plants. Artificial seeds can be used for vegetative species as well as for spermatophytes. In this method, plant cells are cultured to produce a cell lump called embryogenic callus, which divides and is enclosed in alginate gel or other materials where the contents are treated as seeds. Also available is the technique of tank culture which uses plant cells and tissues to mass-produce useful substances. This method whereby only metabolites are produced, is based on a concept similar to that of microorganism culture. One of the examples is the production of shikonin by tank culture of the root cells of *Lithospermum erythrorhizon* which is used for the commercial production of lipstick (rouge) and soap by the private sector as biocosmetics. Other well-known products in the field of foodstuffs include the red dye carthamin produced by safflower cell culture and the yellow dye crocin obtained from the cell culture of *Gardenia augusta* Merr. var. *grandiflora*.

Cell culture and tissue culture are basic biotechnology methods which must be used when plants are produced from cells. Cell fusion whereby combinations of cells which do not occur in nature are fused is also an extremely important technique. In the case of plant cells, external cell walls are lysed by enzymes, and cells become naked cells (protoplasts) which are fused. Cell fusion has led to the production of pomato, a fusion between tomato and potato. Pomato was a symbol of the potential of biotechnology in its initial phase of development but was not a useful plant. The scope of this technology has expanded from simple cell fusion to combination with asymmetric fusion, DNA introduction and nuclear transfer.

Examples of successful cell fusion of citrus fruits include the development of "Oretachi" (orange and trifoliate orange), "Shuvel" (Citrus unshiu and navel) and "Gravel" (grapefruit and navel). Cell fusion has also been actively used for tomato breeding. Another development is the production of "bio-hakuran" by cell fusion between red cabbage and Chinese cabbage.

Recombinant DNA technology which has made rapid progress in recent years, generates a great variety of new plants. Progress in this area can be considered to be the mainstay of plant biotechnology. New development of biotechnology is currently centered on transgenic plants, for which 2,400 cases or more have been field-tested in the United States, Belgium, France and other countries; in Japan, field tests have already been completed for tomato, rice and petunia. Such properties as virus resistance, chemical tolerance and insect resistance are added in the great majority of the cases at present. A large number of products with introduced genes, which are superior in added value or in processing quality for perishable foods, will be developed in the future: for example, a transgenic tomato (Flavr Savr), where the activity of polygalacturonase is controlled at the gene level by antisense RNA, and a hard consistency is maintained without softening even after ripening. Furthermore, genes which promote starch synthesis also will be introduced in these future products.

The Ministry of Agriculture, Forestry and Fisheries initiated the rice genome analysis project in the fiscal year 1991. It will be interesting to determine how far the analysis of genes or mapping work carried out on a global scale will proceed in the 21st century. If mapping work were to involve a majority of plants, conventional genetic phenomena could be analyzed on a molecular basis, and the results used for breeding programs. If the position and time where introduced genes are expressed could be regulated, it will be possible to produce enzymes, hormones and other biologically active substances using rice and other plants. It will also be possible to produce crops rich in special proteins, amino acids, vitamins and sugars. This method is called molecular farming; the studies in this direction are making steady progress.

In the field of fishery-related biotechnology, the growth hormone of such fish as salmon, yellowtail, sea bream and tuna is subjected to recombinant DNA technology, and cultured in large amounts using *Escherichia coli*; the hormone is then administered to the cultured fish. These studies which are in the experimental phase are conducted with a view to improving growth and feed-cost efficiency. The fish exhibit a marked reaction to the growth hormone administered from outside, even if endogenic growth hormone is secreted by their hypophysis.

FOOD BIOTECHNOLOGY

Biotechnology in the field of foodstuffs is centered on the use of microorganisms and enzymes.[4,5] Advanced technology will have a major impact on this field. The products already available on the market as foods are mainly related to carbohydrates. The representative biological functions and types are outlined as follows:

1. Prevention of obesity: Production of low-calorie or noncalorie sweeteners, including sugar alcohols such as maltitol and erythritol.

2. Resistance against tooth decay: These products do not form polysaccharides (dextran and levan causing dental plaques) by *Streptococcus mutans* which is a common microorganism in the oral cavity, and can not be easily changed into organic acid by anaerobic fermentation. They consist of palatinose, fructooligosaccharides and coupling sugar.

3. Superior function for adjustment of intestinal microflora: After the product is taken into the body, it reaches the intestine without being metabolized by the enzymes in the human body, and contributes to the proliferation of *Bifidobacterium*. Among the intestinal bacteria, *Bifidobacterium* is known to be a beneficial intestinal microorganism. Fructooligosaccharides, isomaltose and galactooligosacharides display this function.

4. Maintenance of flavor and improvement of food properties: The products include cyclodextrins, xylobiose and agarooligosaccharides, and are used for the improvement of food properties.

Japan is remarkably advanced in the application of enzyme techniques in the field of carbohydrates and is leading the world in this particular field of research. Because this field will become essential for glyco-engineering, the next generation biotechnology, it is currently drawing the attention of the scientific community worldwide.

Studies in the field of microorganisms related to foods have made progress in the area of cell fusion for improvement of yeast and *Aspergillus* spp. used for the fermentation of refined sake, shochu (distilled spirit) and wine. The studies have contributed to the unique characteristics of the taste and flavor of these products. Furthermore, cell fusion is used to grow the yeast for bread-making which does not assimilate fructooligosaccharides and acts in the adjustment of the intestinal microflora. The resulting bread is produced and sold on the market in which fructooligosaccharides added remain inside without being decomposed during the bread-baking process.

In the production of food materials, cell fusion for the development of microorganisms for amino acid fermentation has been carried out for a long time. Highly efficient strains for the fermentation of amino acids such as lysine, threonine and glutamic acid are being produced.

In addition, studies for the development of mushrooms which have great potential are another important activity.

Amino-acid fermentation and yeast production by genetically modified microorganisms are currently being promoted. These techniques will be extensively used in the field of food processing. A bioreactor which reacts in abnormal environments such as under high temperature and strong acidity conditions will be effectively employed in the food processing industry, as a result of the progress in the field of protein engineering to convert enzyme functions.

A remarkable development is taking place in the large-sized bioreactor for the production of isomerized glucose syrup (High Fructose Corn Syrup). Extensive studies are currently being conducted on its application to the ester-exchange reaction of oil and fat, various oligosaccharides, soy sauce, refined sake, edible vinegar, beer, oligosaccharides and sugar alcohol. These achievements are largely due to the efforts of the Food Industry Bioreactor Research Association under the supervision of the Ministry of Agriculture, Forestry and Fisheries during the period 1984-89. Though enzyme reactions are currently used in limited areas such as oligosaccharides, their applications will expand by the advances in research on enzyme engineering, protein engineering and glyco-engineering, and will be closely related to the development of novel biologically functional foodstuffs.

COMMERCIALIZATION IN THE USE OF GENETICALLY MODIFIED ORGANISMS

SAFETY AND GUIDELINES FOR BIOTECHNOLOGY[6,7]

For the application of new techniques in the field of biotechnology, guidelines on recombinant DNA (rDNA) techniques in particular have been issued one after another following the Asilomar Conference held in 1975. NIH guidelines were enacted in the United States in 1976, along with the recombinant DNA experimental guidelines in Britain in the same year, followed by France in 1977 and in Germany in 1978. In Japan, the recombinant DNA experiment guidelines for universities were issued by the Minister of Education in 1979 and for governmental research institutes and the private sector by the Science and Technology Agency in the same year. These guidelines are issued to guarantee safety by a combination of two measures: physical containment to ensure that genetically modified organisms will not go out of the laboratory and biological containment to ensure that they will not continue to live in the general environment.

However, several years after the promulgation of these guidelines, repeated recombinant DNA experiments have revealed that there is no such danger as was anticipated in the initial stage. Thus, deregulation of standards for such hosts as *Escherichia coli* and limited vectors has been promoted several times so far. Due to the development of the recombinant DNA technologies, plants, animals and foods produced by these technologies have reached the phase of commercialization. Establishment of guidelines for industrialization has proceeded in contrast to guidelines for experiments.

As a global move, biotechnology safety and regulations were taken up by OECD in 1983. "Recommendation on Safety Problems in The Industrial Use of Genetically Modified Organisms" was issued in 1986. "Good Development Standards for Small-Scale Field Test of Genetically Modified Organisms" and "Good Industrial Large Scale Practice for Tank Culture of Genetically Modified Organisms" were submitted in 1990. "Study of Large Scale Field Test," "Study on The Safety of Biotechnology-Based Foods," and "Genetically Modified Organisms Monitoring Technique in The Environment" were reported in 1991.

Following these studies, "rDNA Technology Industrialization Guidelines" were issued by the Ministry of International Trade and Industry in Japan in 1986. "Guidelines on the Production of rDNA Technology-Based Medical and Pharmaceutical Products" were issued by the Ministry of Health and Welfare in the same year. The Ministry of Agriculture, Forestry and Fisheries developed "Guidelines for the Use of Genetically Modified Organisms in the Field of Agriculture, Forestry and Fisheries" in 1989. In Japan, industrialization of recombinant DNA technologies proceeds with a view to enforcing the compliance with these guidelines. In 1991, "rDNA-Based Food and Food Additive Production Guidelines" and "Safety Evaluation Guidelines for rDNA-Based Foods and Food Additives" were set up by the Ministry of Health and Welfare. At this stage, foods containing genetically modified organisms were not included in the guidelines.

The number of field tests conducted is still limited in Japan. Isolation field tests have so far been carried out or are being pursued for the following:

1. TMV-resistant tomato plants (Japan's first genetically modified plants) tested in the isolation field of the National Institute of Agro-Environmental Sciences, MAFF in 1991;
2. Stripe virus-resistant rice in 1993;
3. CMV-resistant melon and CMV-resistant petunia in 1993;
4. CMV-resistant tobacco in 1994; and
5. Rice with low-glutelin content and low allergen in 1994.

The following have been or are still being subjected to ordinary greenhouse experiments which precede the isolation field tests;

1. Dwarf Turkish platycodon in 1992;
2. Rice with a low amylose content and PLRV-resistant tomato in 1993; and
3. CMV-resistant tomato and herbicide-resistant soybean in 1994.

Agricultural research aims at new objectives such as promotion of agriculture friendly to the environment and crops with outstanding properties for processing adaptability, in addition to the conventional target of high yield and disease resistance. In this sense, early commercialization of biotechnological achievements is eagerly anticipated.

In May and October 1993, the author had the opportunity to visit some U.S. companies which lead the world in the field of plant biotechnology, and administrative organizations in charge of regulations to discuss various problems relating to the commercialization of biotechnology. He was impressed by the collection of a vast amount of data demonstrating perfect safety.

One of the subsequent tasks is to make efforts to win the understanding of the general public on biotechnology safety; this is as important as the advances and development of science and technology. Further approaches to commercialization require the acceptance of new technologies by the general public.

The Ministry of Agriculture, Forestry and Fisheries is currently implementing the research project entitled: "Development of Assessment Techniques for Effective Use of Genetically Modified Organisms" based on close cooperation among the industrial sector, academic circles and government. This project will require special consideration of government policies.

REFERENCES

1. Kainuma K. JIRCAS Newsletter 1993; 1:2-3.
2. Kainuma K. Advances and safety of biotechnology. In: Namiki S, ed. White paper on food 1993 "Safety of food and agricultural products". Tokyo: Food Agriculture Policy Research Center, 1994:181-195.
3. Kainuma K. Present status of food biotechnology and prospect of 21st century. In: Kainuma K, ed. Food Industry and Biotechnology. Tokyo: Meibun Shobo, 1990:137-149.
4. Kainuma K, Hidaka H, Sasaki T. Present situation and future prospect of enzyme utilization. In: Kainuma K, ed. Food Biotechnology. Tokyo: Meibun Shobo, 1988.
5. Kainuma K. 21st century, technology will change your table. Newton 1989; 9:96-101.
6. Hasebe R. Safety of agricultural, food biotechnology. In: Uchiyama M, ed. Encyclopedia of Safety Supply of Food. Tokyo: Sun Choh, 1994: 306-310.

Agricultural Biotechnology for Sustainable Productivity (ABSP)
A U.S. Agency for International Development (USAID) Initiative

John H. Dodds

INTRODUCTION

The ABSP project is part of an on-going commitment of USAID to plant biotechnology. One of the initial U.S. Agency for International Development (AID) centrally-funded initiatives in this area was the Tissue Culture for Crops Project (TCCP) based at Colorado State University (CSU) in Fort Collins, Colorado, U.S., which sought to produce crops (wheat, rice and sorghum) tolerant to an array of stresses, including salinity, drought and acid/aluminum soil conditions. Some of the most significant accomplishments from TCCP were the registration and release of sorghum germplasm with improved tolerance to fall armyworm and to acid/aluminum soil conditions (other work on salt and drought tolerance is ongoing) and the construction of the International Plant Biotechnology Network (IPBNet).[1]

In 1989, the USAID Office of Agriculture began a multi-stage review of opportunities to support biotechnology. External evaluation

Plant Biotechnology Transfer to Developing Countries,
edited by D.W. Altman and K.N. Watanabe. © 1995 R.G. Landes Company.

of TCCP was undertaken, followed by the convening of an expert panel under the direction of the National Research Council (NRC). This panel produced a report, "Plant Biotechnology Research for Developing Countries," published by the NRC in 1990,[2] which looked at constraints on productivity in the developing world and relevant technologies to address those constraints in the near future.

The combined results of those investigations were then reviewed by AID missions in developing countries and national and international agricultural programs. A new project in plant biotechnology was then designed which would bring together public sector and commercial research efforts in an integrated product-development program.

AGRICULTURAL BIOTECHNOLOGY FOR SUSTAINABLE PRODUCTIVITY (ABSP): A NEW FOCUS, A NEW APPROACH

The purpose of the Agricultural Biotechnology for Sustainable Productivity (ABSP) project is "to mutually enhance U.S. and developing country institutional capacity for the use and management of biotechnology research in developing environmentally-compatible, improved germplasm, through exchanges and training courses with developing country scientists."

MANAGEMENT APPROACH

Unlike many basic research programs, ABSP takes an integrated approach to the development of specific research products and their transfer to developing-country partners. In order to achieve integration, ABSP has set for itself the following management guidelines:

- Maintain a highly-focused research program concentrating on specific crops and technologies.
- Implement a product-oriented research style which links public- and private-sector institutions in the U.S. and developing countries.
- Link product-oriented research to policy analysis of intellectual property and biosafety to ensure product commercialization in an environmentally and socially responsible manner.
- Maintain a geographical focus on specific centers of expertise; develop a critical mass for the multi-disciplinary team which will transfer technology to national programs.
- Build a global network that provides access to information for developing and developed countries worldwide and serve as a forum for the exchange of ideas and information on biotechnology in relation to sustainable agriculture.
- Establish linkages with other organizations such as the CGIAR Agricultural Research System and the Biotechnology Industry Organization (BIO).

RESEARCH FOCUS

The goal of the ABSP project research is to cooperate with developing countries in adopting a wider application of biotechnology in order to address priority problems which represent specific constraints on agricultural productivity. ABSP is working with developing-country scientists to genetically-engineer pest and pathogen resistance, ABSP focuses on the potential to reduce chemical inputs and produce high-quality plant material at a lower economic and environmental cost than conventional methods alone. The integration of new technologies, such as plant genetic transformation and bioreactor micropropagation, into the mainstream of international agriculture is envisioned. In achieving this goal, the economic and environmental sustainability of agricultural production systems will be improved and the quality of life enhanced by increasing the availability of food for consumption and marketing.

The research problems identified for inclusion in ABSP are those which conventional plant breeding alone cannot resolve. From the three regions of collaboration, Asia, Africa, and Latin America, crops were chosen based on economic and nutritional significance coupled with severe pest or pathogen constraints on productivity. Targeted crops and constraints include: potato, constrained by potato tuber moth; sweet potato, constrained by sweet potato weevil; maize, constrained by the maize stem borer; cucurbits, constrained by zucchini yellow mosaic virus and other cucurbit potyviruses; and tomato, constrained by tomato yellow leaf curl virus and/or beet curly top virus.

ABSP has set research objectives which, when considered in their entirety, will reduce the constraints on productivity and broaden the application of biotechnology in developing countries:

- Transfer scientific knowledge and techniques to developing countries through postdoctoral fellowships. The program in its total has about 14 postdoctoral scientists involved.
- Assemble minigenes containing insect-resistance genes (Bt and proteinase inhibitor), driven by plant-specific regulatory elements. This work builds onto ongoing expertise by Prof. Ray Wu and his cooperators.[3,4]
- Genetically engineer potato, sweet potato, and maize for resistance to insect pests in developing countries. This work builds on expertise developed by Dr. Dave Douches, Dr. Jan Tippet and Dr. Mariam Sticklen and other cooperators, in the areas of regeneration and transformation technology.[5-9]
- Genetically engineer cucurbits with a virus coat-protein gene for development of resistance to potyviruses. This follows ongoing work by Dr. Rebecca Grumet and her cooperators.[10,11]
- Genetically engineer tomato for resistance to viruses causing tomato yellow leaf curl disease. This links into the ABSP program specific ongoing research by Prof. Roger Beachy and his co-workers.[12]

- Demonstrate pest resistance of transgenic crops at the laboratory, greenhouse and field level, and integrate this into sustainable agricultural systems via collaborations with plant breeders, agronomists, statisticians, virologists and entomologists (with expertise in insect-resistance management and integrated pest management).
- Transfer DNA Plant Technology bioreactor-micropropagation technology to private-sector collaborators for the propagation of banana, pineapple, coffee and ornamental palm. This work is led by Dr. Neal Courtenay Gutterson.

THE U.S. PUBLIC SECTOR RESEARCH TEAM

ABSP is the first and only comprehensive biotechnology program to utilize collaborative research teams which encompass the public and private sectors, field agronomists, breeders, entomologists, virologists, molecular biologists, experts in biosafety and intellectual property protection and other specialists committed to the equitable use of biotechnology in agricultural research, globally. These teams provide the flexibility to implement partnerships in commercially-oriented research as easily as research at land-grant institutions, and through national and international agricultural research centers.

A consortium of three U.S. universities (Michigan State University, Cornell University and Texas A&M University) has developed a team of scientists with considerable expertise, who will lead the research and training components set forth in the project objectives.

Developing-country maize genotypes will be improved through plasmids constructed by Dr. Ray Wu, biochemist, molecular geneticist and faculty member in the Section of Biochemistry, Molecular and Cell Biology at Cornell University, Ithaca, New York, U.S. Dr. Wu has focused on isolating and characterizing cereal regulatory elements and using those elements to increase the expression of foreign genes in cereal crops. These genes were used by Dr. M. Stricklen at Michigan State.

Systems to transform plants from shoot apices using *Agrobacterium* have been developed by Dr. Roberta Smith, the Eugene Butler Professor in the Department of Soil and Crop Sciences at Texas A&M University, College Station, Texas, U.S.

Transformation systems for cucurbit crops, that are of significant value as cash crops to many developing country producers, have been developed by Dr. Rebecca Grumet, molecular virologist and faculty member in the Department of Horticulture at Michigan State University, who has cloned viral coat proteins and transferred those genes into crop species.

The potato research program is coordinated by Dr. Dave Douches, potato geneticist, breeder and faculty member in the Department of Crop and Soil Sciences at Michigan State University, is carrying out research using Bt genes to transform potato germplasm, adapted to conditions in North Africa and the Middle East, for resistance to potato tuber moth.

With the recent implementation of a new cooperative agreement between ABSP and the Agricultural Genetic Engineering Research Institute (AGERI) in Cairo, Egypt, a number of collaborative research activities have resulted in the addition of Drs. Roger Beachy and Claude Fauquet of the Scripps Research Institute to the research team, who will investigate the causal agent in tomato yellow leaf curl disease. Also, Drs. Henry Munger, Rosario Provvidenti and Molly Kyle of Cornell University will team with Rebecca Grumet to develop transgenic cucurbits resistant to potyviruses. Dr. Ed Grafius, from the MSU Entomology Department will team with Dave Douches to develop transgenic potato germplasm resistant to potato tuber moth.

PUBLIC/PRIVATE SECTOR RELATIONSHIPS

Recently, many AID foreign-assistance programs have begun to foster a more active involvement on the part of private-sector entities in the U.S. and its client countries. This involvement has particular benefit for programs in biotechnology because much of the technology is based in the private sector. AID and its clients have a continuing interest in realizing the full scientific and institutional benefit of past investments in agricultural research.

The ABSP project fosters direct linkages between the public and private sectors including the U.S. university community (Michigan State University, Texas A&M, Cornell University and University of Arizona) and other public-sector institutions (Scripps Institute, La Jolla, California, U.S.) as well as the private sector (DNA Plant Technology, Cinnaminson, New Jersey, U.S. and ICI Seeds, Inc. Slater, Iowa, U.S.). Linkages with developing-country institutions also include a variety of public institutions (KARI, Nairobi, Kenya; CRIFC, Bogor, Indonesia; University of Costa Rica; and AGERI, Cairo, Egypt) and the private sector (Agribiotecnología de Costa Rica, Alajuela, Costa Rica and Fitotek Unggul, Jakarta, Indonesia).

ABSP is attractive to developing countries which are seeking to move into the biotechnology arena but are unable to access technology which is quickly becoming more proprietary and "privatized." Private-sector linkages have been established in ABSP from its initiation, so involvement with the project gives developing-country programs direct access, which can serve to drive reform within their own research system.

As an example, DNA Plant Technology Corporation (DNAP), has established two joint projects with private biotechnology companies under the ABSP project, which are led by Dr. Neal Courtney Gutterson, Tropical Crops Research Director at DNAP. The first joint project is with Agribiotecnología de Costa Rica S.A., to explore advanced micropropagation methods (bioreactor cloning) for banana, pineapple, coffee and ornamental palms. The second involves a micropropagation company, Fitotek Unggul, Jarkarta, Indonesia, and focuses on pineapple micropropagation.

A collaborative ABSP project between the Central Research Institute for Food Crops (CRIFC), Bogor, Indonesia and ICI Seeds, Inc., Slater, Iowa, U.S., which will develop insect-resistant tropical maize for Indonesia, through genetic engineering, is led in the U.S. by Dr. Martin Wilson, Cell Biologist Project Leader at ICI Seeds. CRIFC is a public sector, developing-country ABSP partner and ICI Seeds is a U.S. private-sector partner.

The immediate goals of the 3-year CRIFC/ICI collaboration are (a) to produce commercially-important insect-resistant maize germplasm for Indonesia and (b) to work with Indonesian scientists in the genetic engineering-enabling technology. In the longer term, it is hoped that commercialization of the germplasm can be achieved through partnership with a private company in Indonesia.

NETWORKING

The ABSP network publishes a quarterly newsletter, Bio*Link*, which provides access to literature, builds connections and promotes interaction among institutions and individuals through print and electronic media, organizes country-specific, regional and global workshops, conferences and symposia, coordinates internship programs in biosafety and intellectual property rights and facilitates the building of linkages to other USAID projects such as the Bean/Cowpea CRSP as well as the International Agriculture Research Centers and other institutions involved in international agricultural-biotechnology research and development.

ABSP fosters an integrated approach to the promotion of biotechnology in developing countries by supporting human resource development and technology transfer through postdoctoral fellowships in research areas and through internships in technical and policy areas, such as intellectual property and biosafety, which may have a direct impact on the success and adoption of the technology. By supporting consultants and developing country interns in biosafety and IPR, ABSP hopes to establish a policy environment which encourages the growth and commercial potential of biotechnology. This support is important to developing-country programs which recognize that increased capability in these areas is critical to the success of fledgling biotechnology programs.

INTELLECTUAL PROPERTY RIGHTS (IPR)

Licensing and intellectual-property issues that arise in connection with the transfer to developing nations of technologies developed under collaborative programs are addressed by John Barton, George E. Osborne Professor of Law at Stanford Law School, Stanford, California, USA. He works with developing-country officials and ABSP cooperators to assist in the drafting of international agreements between developed- and developing-country affiliates. The ABSP management team works with developing-country institutional leaders to identify candidates to participate as interns, under the direction of Mr. Barton, in intellectual-property training programs. The program provides interns with access to expertise and information regarding IPR, not only from a legal perspective but also from the business perspective of private biotechnology companies. Interns receive hands-on training and develop case studies based upon current and expected situations in their home countries.

BIOSAFETY

The United States Department of Agriculture, Animal and Plant Health Inspection Service (APHIS), Biotechnology, Biologics, and Environmental Protection (BBEP) division, Washington, DC, U.S., is represented on the ABSP Technical Advisory Group by Dr. Sivramiah Shantharam. Scientific, technical, and regulatory advice on matters related to biosafety, environmental safety, and the exportation and importation of genetically engineered organisms is provided by Dr. Shantharam through consultation on an institutional basis and through individual interaction as part of a biosafety internship program.

Biosafety interns from developing countries are trained at Michigan State University and ICI Seeds, Inc., an ABSP corporate collaborator, in the safe handling of transgenic materials in the laboratory, the greenhouse and the field.

ASSOCIATION OF BIOTECHNOLOGY COMPANIES/BIO

The ABSP commitment to building diverse linkages is exemplified in the project's role as sponsor of memberships in the Biotechnology Industry Organizations (BIO), Washington, DC, U.S. Developing-country and U.S. institutions collaborating in ABSP, both public and private, were originally provided membership in the Association of Biotechnology Companies (ABC), a not-for-profit trade association, formed in 1983, with more than 350 members from 29 countries. It had as its members companies, universities and research institutions throughout the world.

INDUSTRIAL SEMINAR SERIES

Government, private-sector and public-institution leaders from ABSP focus countries have participated in a prototype Industrial Seminar Series. The purpose of the seminar series is to expose leaders in biotechnology research and development from countries to U.S. companies with agricultural plant-biotechnology programs in order to interact with technical and business specialists responsible for the implementation and incorporation of those programs into the corporate structure. The prototype series met with enthusiastic reviews by participants and a future series is being planned.

PUBLICATIONS

Bio*Link* is distributed to 113 countries and over 2,000 individuals and institutions; approximately half of that distribution is to developing countries. The cost of producing the newsletter is covered by funding within the project, so there is no subscription fee. New subscribers to Bio*Link* from around the world are always welcome.

ACKNOWLEDGEMENTS

The ABSP Project is truly a research team. The original plan was to include all people as co-authors. Given that this cannot be done for logistical reasons, I would like to acknowledge the following cooperators: Judy Chambers, John Barton, Roger Beachy, David Douches, Claude Fauquet, Ed Grafius, Rebecca Grumet, Molly Kyle, Henry Munger, Rosario Provvidenti, Sivramiah Shantharam, Roberta Smith, Maro Söndahl, Mariam Sticklen, Rick Ward, Martin Wilson, Ray Wu, Bruce Bedford, Karim Maredia, Cindy Jennings, Dean Norton, and Linda Skiba. We also acknowledge financial support from the U.S. Agency for International Development (USAID) under cooperative agreements nos. DAN 4197-A-00-1126-1126-00, LAG 4197-A-00-2032-00 and 263-0152-A-00-3036-00.

REFERENCES

1. IBPNet Membership Directory. Fort Collins: Colorado State University Press, 1988.
2. National Research Council. Plant Biotechnology Research for Developing Countries. Washington, DC: National Research Council, 1990.
3. Wu R, Lin E, Kung SD. Proc. of Conference on Genetic Engineering and Biotechnology. Beijing: Beijing University, 1984.
4. Wu R, Grossman L, Moldave K. Recombinant DNA Methodology. New York: Academic Press, 1989.
5. Sticklen MB. Genetic engineering of plants: An alternative to pesticides and a new component of integrated pest management. In: Weigmann D, ed. Pesticides for the Next Decade: The challenges Ahead. Proc. the Third National Research Conference on Pesticides. Nov. 8-9, 1990: 552-566.
6. Sticklen MB, Hajela RK, Hajela N, McElroy D, Cao J, Wu R. Polygenic transformation of rice using a microprojectile delivery of DNA. In: Swaminathan MS, Oka HI, Khush GS, Futsuhara Y. eds. Proc. Second International Rice Genetics Symposium. Los Baños, Philippines. May 14-18, 1990.
7. Zhong H, Srinivasan C, Sticklen MB. In-vitro morphogenesis of corn (*Zea mays* L). I. Differentiation of multiple shoot clumps and somatic embryos from shoot tips. Planta 1992; 187:483-489.
8. Zhong H, Srinivasan C, Sticklen MB. In-vitro morphogenesis of corn (*Zea mays* L). II. Differentiation of ear and tassel clusters from cultured shoot apices and immature inflorescences. Planta 1992; 187:490-497.

9. Cheng J, Bolyard MG, Saxena RC, Sticklen MB. Production of insect resistant potato by genetic transformation with a δ-endotoxin gene from *Bacillus thuringiensis* var. *kurstaki*. Plant Science 1992; 81:83-91.

10. Grumet R. Genetically engineered plant virus resistance. Hortscience 1990; 25:508-513.

11. Grumet R, Fang G. cDNA cloning and sequence analysis for the 3' terminal region of zucchini yellow mosaicvirus RNA. J Gen Virol 1990; 71:1619-1622.

12. Beachy RN, Loesch-Fries S, Tumer NE. Coat protein mediated resistance against virus infection. Ann Rev Pathol 1990; 28:451-474.

CHAPTER 18

MODERN BIOTECHNOLOGY AT INTERNATIONAL AGRICULTURAL RESEARCH CENTERS

N.C. Brady

Genetic improvement was the driving force behind the Green Revolution. The development, dissemination and use of high-yielding varieties of major food crops that responded to irrigation water and chemical inputs provided farmers in low-income countries with tools they had never had before. They responded by increasing food production at rates even greater than increases in human population numbers. As a result, per capita food production increased in all areas of the world, except for SubSaharan Africa, and food prices generally declined. The poorest of the poor were the primary beneficiaries since the Green Revolution helped assure them of a steady supply of low-cost food.

A primary factor in the process of genetic improvement was the creation in the 1960s and 1970s of some 13 international agricultural research centers (IARCs), located mostly in Third World countries. Under the auspices of a multi-donor Consultative Group on International Agricultural Research (CGIAR), ten of these centers focused on modifying the genetics of crop plants and animals. In turn, these IARCs cooperated with dozens of national centers in Third World countries and helped them organize and implement viable agricultural research endeavors.

From the very inception of the first of these IARCs, research was initiated to improve the genetic ability of selected crop plants to produce high yields, and to withstand the stresses such as insect and disease pressures, drought, soil acidity and salinity and low and high temperatures. The researchers knew that the small farmers they served needed to have production tools adapted to the environment in which they lived. They

Plant Biotechnology Transfer to Developing Countries,
edited by D.W. Altman and K.N. Watanabe. © 1995 R.G. Landes Company.

could not afford the inputs of chemical pesticides which at that time were perceived as being essential for crop protection. They could not provide the limestone needed to neutralize high soil acidity and aluminum toxicity, nor could they properly drain low-lying but productive soils. Only with early government subsidies for irrigation water and chemical fertilizers were they able to take advantage of these two important inputs.

These small farmers were poor and generally uneducated. Complex inputs and management practices were not easy for these illiterate men and women to comprehend. Researchers assumed that if most of the desired technology could be associated with the seeds the farmers used, the small farmers would benefit as much as the larger ones from the new technologies.

This assumption proved to be largely viable as the first products began to appear from these genetic-improvement scientists. Greatest success of the system was experienced initially with the genetic improvement of the major cereals wheat, rice and maize using conventional plant breeding techniques. The process consisted of collecting genetic resources of these cereal crops from around the world, storing them in modern gene banks and characterizing them for significant agronomic, grain quality and pest-resistant features. By appropriate crossing and selection, and by widespread field assessment of the progeny obtained, new production tools were made accessible to farmers around the world, rich and poor. Hundreds of scientists from developing countries joined in collaborative research endeavors that brought about unprecedented improvements in crop productivity.

These remarkable changes were soon evident in the farmers' fields, and in the national food production statistics of the Third World, especially in Asia. Unparalleled increases in the production of wheat and rice took place in the 1970s. These increases outstriped population increases, although the latter were also unprecedented. The Green Revolution was in full swing.

Unfortunately, early in the game, constraints were noted in this comprehensive approach to genetic improvement. First, even among the tens of thousands of cultivars collected for the gene banks, scientists did not find accessions that accommodated or controlled some of the known major crop-production constraints. For example, none of the accessions were found to be resistant to attacks of some insect pests and diseases.

Second, it was not always possible to incorporate genes of the germplasm-bank accessions into existing cultivated varieties. This was true especially where the desired genes were found in wild plant types or species different from those being cultivated. Among rice varieties, such difficulty was sometimes encountered between the indica-type rices of the tropics and their temperate-zone japonica counterparts.

Thirdly, even where successful gene incorporations did take place through traditional breeding methods, the resulting progeny did not always provide sustainable production. Vertical resistance of the newly-created varieties to insects and diseases soon "broke down," often within only 3-4 years of the variety's being released to farmers. This forced the plant breeders and farmers to be on a treadmill, barely able to stay ahead of the mutating insect pests and pathogens.

This brings up the fourth constraint, which is the time consumed in conventional plant breeding. The process of crossing, selection and evaluation, dissemination and seed increase is a lengthy one, often consuming from 6 to 10 years from the time of the first cross to the ultimate extensive use of new varieties in farmers' fields. Breeders have sought techniques that would shorten this time requirement and reduce the labor consumed in creating and evaluating a new cultivar.

These difficulties led researchers to seek tools that would help them overcome these production constraints. One such tool is that of tissue culture. A few examples of the use of this tool will be given. By the mid 1970s, tissue-culture techniques were being used over wide areas of mainland China to try to improve rice varieties. Numerous lighted greenhouses and converted storage areas were used by researchers in communes throughout the rice growing areas to culture rice tissues and to regenerate them into green plants. Through what seemed to be trial and error techniques, some of these tissue culture researchers were successful and many new varieties were made available for farmer use.[1]

Starting in the mid 1970s, anther culture and isolated-microspore culture were used at the International Rice Research Institute (IRRI). Considerable progress has been made in replicating for tropical-indica rices the tissue-culture successes achieved with temperate-zone japonica rices.[2,3] Initial efforts to use somaclonal variation to enhance tolerance of indica rices to stresses such as salinity and cold tolerance met with only modest success.[4] However, anther-culture techniques have been found that have reduced by about half the time required to provide homozygous diploid lines.[5] Hundreds of plants have been generated from F_1 crosses and the progeny evaluated through the world-wide International Network for Genetic Evaluation of Rice (INGER).

In recent years, research at IRRI has focused on the plant regeneration from protoplasts of indica rices and in optimizing conditions for rice transformation and co-transformation using plasmids carrying either a marker and/or a selectable gene (F.J. Zapata, personal communication).

Tissue culture has also been a profitable tool use by the International Potato Center (CIP), headquartered in Peru. Tissue culture techniques have been utilized since the early 1970s to enhance in vitro conservation and exchange of potato genetic resources and to assure disease-free potato stocks for this international exchange.[6-8] Likewise, dependable ELISA field kits were developed to detect potato leaf roll virus and potato virus Y. These kits are in commercial use in some potato-growing regions of the Third World.[9]

In carrying out much of its tissue- and cell-culture research, CIP utilized collaborative research arrangements with cooperators in universities and research centers in the more advanced countries.[10] This networking approach has been very helpful as CIP has begun to make use of other biotechnology inputs, including genetic engineering.

The International Institute for Tropical Agriculture (IITA) has used tissue culture to create and disseminate virus-free cassava plantlets for international exchange.[11] IITA is cooperating with advanced laboratories in anther and pollen culture as well as more sophisticated research approaches.

The tissue-culture research has led most of the IARCs concerned with plants and animals to the use of more sophisticated molecular-biology approaches. While numerous products from this research have not yet materialized, the possibilities for future accomplishments abound. The crop-oriented centers generally have biotechnology research units focusing on endeavors such as gene mapping, marker-assisted selection for breeding and gene transfer. Much of this research is being done in cooperation with more advanced laboratories in Europe, the United States or Japan. There has been a decided upswing in such biotechnology in the IARCs and their national counterparts. In cooperation with scientists in both the North and the South, IRRI is helping to integrate the development of an RFLP map for rice.[12] IRRI has successfully transformed to rice the agronomically-important soybean trypsin-inhibitor gene that confers resistance to the yellow stemborer.[3] IRRI has also helped develop methods for the regeneration of some indica rice protoplasts including some whose transformation has taken place (Wu et al, in press).

IITA has begun genetic-engineering research on bananas, cowpeas and cassava. Genetic maps using RFLPs are being developed in cooperation with overseas institutions.[13] The first stable genetic transformation of the cowpea came from such overseas collaboration.[1] CIP is using genetic engineering for transformation of the potato through the insertion of antibacterial protein genes into potato clones. Potato gene mapping is being pursued,[14] much of it in cooperation with scientists in the United States and Europe.[15]

Table 18.1 illustrates the degree to which various biotechnology techniques are being used at the crop-oriented IARCs. In recent years, research leading to transformation has been undertaken by several of the centers. This has been made possible by breakthroughs in more advanced laboratories of industrialized countries that have provided techniques that the centers can apply in improving their mandate crop.

Linkages with these more advanced laboratories are commonly leading to the creation of informal or formal networks in which the centers are a part. In 1985, the Rockefeller Foundation established a rice biotechnology network that has set a pattern which is being closely observed by the CGIAR system.[16] Laboratories in both the North and South (including IRRI) are focusing on a variety of research endeavors to bring about transformations for rice and to provide breeders with genetic materials to combat production constraints.

Table 18.1. Application of biotechnology for commodity crops of the CGIAR centers

Biotech Activity	International Centers								
	CIAT	CIP	CIMMYT	ICARDA	IPGRI	IITA	IRRI	ICRISAT	INIBAP
Anther/ microspore culture	1	2						1	
Embryo rescue	1/2	1[a]				1/2	2	1	1
Electrophoresis	1/2	1/2	1	1					
ELISA		1							1
In vitro plantlets		1							
In vitro conservation	1/2	1[a]/2[a]			2	1/2	2		
Meristem culture/ Thermo-therapy		1/2							
Micropropagation		1/2				2	2		
Monoclonal Antibody		2	1			2		1	
Nucleic Acid diagnostic probe		1[a]/2						1	
Protoplast fusion		2						1	
RFLP mapping	1[a]/2	1[a]/2	2[a]				1/2		
Somatic cell culture	2	2[a]				2			
Somaclonal variation		2	2		2	2	1		
Shoot tip culture Thermo-therapy	1	2[a]							
Tissue culture (alien genes)			2						
Transformation	2	1[a]/2	2[a]			2		1/2[d]	2[d]
PCR	1[a]	1[a]	1[a]	1[a]		1[a]	1[a]	1[a]	

1 Research at Center
2 Reserach at Center and with overseas cooperation
a Started since Plucknett (1990) publication
Adapted from Plucknett et al[18]

A similar North-South network pattern was used in 1988 by CIAT in setting up a Cassava Biotechnology Network,[13,17] and by CIP in its genetic engineering and RFLP analyses. CIP has also participated in an Andean Development Bank-stimulated network of Andean countries focusing primarily on the disease-free potato seed stocks.[16] CIMMYT initiated a North-South RFLP network to help map the genome of maize.[18] IRRI has recently announced an Asian Rice Biotechnology Network that will enhance collaboration among rice-growing countries of that region.[19]

The crop-oriented IARCs have moved at a deliberate but not rapid pace in adopting various biotechnology methodologies. However, the International Laboratory for Research on Animal Diseases (ILRAD) began to make use of these technologies soon after it was established in 1973. This center has the mandate of creating the means of managing two devastating animal diseases, theileriosis (East Coast fever) and trypanosomiasis. These diseases are caused by infection with strains of protozoa transmitted by ticks (theileriosis) and tsetse fly (trypanosomiasis). They threaten the lives of 25-45 million cattle in Africa and also infect other animals.

The development of vaccines against these diseases has been a primary goal of ILRAD, and both conventional and modern-biotechnology tools have been employed. Parasite-specific monoclonal antibodies and DNA probes, and the polymerase chain reaction (PCR) technique have been used to distinguish species and stocks of these organisms.[20] These tools have been helpful not only in seeking viable vaccines for these diseases, but in developing reliable field diagnostic kits.

The search for effective vaccines has had some success with theileriosis but the antigenic diversity stemming from different parasite stocks make these vaccines problematic. Immunization against trypanosome populations has been even less successful. This is due to the remarkable ability of the parasite to evolve new antigens in response to antibody attacks.[20] This antigenic-variation defense mechanism has prevented the development of effective vaccines against this disease.

The investment of IARCs in modern biotechnology was about U.S. $14.5 million in 1990, about half of which was expended by ILRAD.[21] While this level of support has likely increased somewhat, overall budget constraints and the desire of the CGIAR to give more attention to natural-resources issues has prevented large increases in resources for biotechnology.

References

1. Oono K. Rice breeding through tissue culture. In. Zakai AH, ed. SABRAO. Kuala Lumpur, 1988:217-223.
2. Cho MS, Zapata FJ. Plant regeneration from isolated microspore of indica rice. Plant Cell Physol 1990; 31:881-885.
3. Zapata FJ. Accomplishment of IRRI tissue culture laboratory 1980-1993. Final Report. Manila: International Rice Research Institute, 1993.

4. Zapata FJ, Aliyar MS, Tarrizo LB, Narero AU, Sigh VP, Senadhira D. Field performance of anther-culture-derived lines from F1 crosses of indica rices under saline and nonsaline conditions. Theor Appl Genet 1991; 83:6-11.

5. Draz AE, Zapata FJ, Khush GS. Development of dihaploid rice lines through anther culture II. IRRI Newsletter 1991; 16:6.

6. Espinoza N et al. Tissue culture: Micropropagation, Conservation and Export of Potato Germplasm. Special Technical Document 1. Lima, Peru: International Potato Center, 1986.

7. Lizarraga R et al. Tissue Culture for Elimination of Pathogens. Special Technical Document 3. Lima, Peru: International Potato Center, 1986.

8. CIP. Profile 1972-2010. Lima, Peru: International Potato Center, 1987.

9. Brovelli T. Fighting potato virus in Argentina. CIP circular 1989; 17(4).

10. Dodds JH. CIP collaborative research on tissue culture and genetic engineering for potato improvement. CIP circular 1987; 15(1).

11. IITA. 1991 Annual Report. Ibadan, Nigeria: International Institute of Tropical Agriculture, 1992.

12. McCouch SR, Tanksley SD. Development and use of restriction fragment length polymorphism in rice breeding and genetics. In: Khush GS, Toenniessen GH, eds. Rice Biotechnology. Manila, Philippines: IRRI, 1991:109-134.

13. Angel F et al. The use of RFLPs and RAPDs in cassava genomic studies. In: Roca GA, Thro AM, eds. Proceedings of the First International Scientific Meeting of the Cassava Biotechnology Network, Cartagena, Colombia, August 25-28, 1992. Cali, Colombia: CIAT, 1993:62-68.

14. Watanabe K, Dodds JH. Tools of molecular biology expand use of CIP genetic resources. CIP circular 1991; 18(2).

15. CIP Annual Report 1991. Lima, Peru: CIP, 1992.

16. Platais KW, Collinson MP. Biotechnology and the developing world. In: Finance and Development, March 1992. Washington, DC: The IMF and the World Bank, 1992:34-36.

17. Roca WM, Thro AM, eds. Proceedings of the First International Scientific Meeting of the Cassava Biotechnology Network, Cartagena, Colombia, August 25-28, 1993. Cali, Colombia: CIAT, 1993.

18. Plucknett DL, Cohen JI, Home ME. Role of international agricultural research centers. In: Persley GJ, ed. Agricultural Biotechnology: Opportunities for International Development. Wallingford, UK: CAB International, 1990; 14-44.

19. IRRI. Germany's GTZ and ADB funding new Asian rice biotech network. Manila, Philippines: IRRI, 1993.

20. ILRAD. 1991 Annual Report. Nairobi, Kenya: ILRAD, 1992.

21. Collinson MO, Platais KW. Biotechnology and the international agricultural research centers of the CGIAR. In: Peters GH, Stanton BF, eds. Sustainable Agricultural Development: The Role of International Cooperation. Dartmouth, UK: International Conference of Agricultural Economists, 1992:355-367.

CHAPTER 19

RESEARCH COLLABORATION, MANAGEMENT AND TECHNOLOGY TRANSFER; MEETING THE NEEDS OF DEVELOPING COUNTRIES

Joel I. Cohen and John Komen

ABSTRACT

The Intermediary Biotechnology Service (IBS) assists national agricultural research systems in developing countries with needs-oriented biotechnology program management. This includes advice on policy formulation, country reviews to identify priority problems amenable to solution through biotechnology and identifying international biotechnology program expertise for collaboration. Activities and publications are developed to facilitate use of this information and to help IBS work with national and international programs in an intermediary manner. Much information obtained and analyzed to date concerns three issues which research leaders in developing countries face regarding biotechnology program implementation. This chapter first introduces IBS and describes its methodology. Then, the three issues identified for discussion are presented with an analysis of each one provided from the information collected and examined by IBS. The issues selected for discussion are: research collaboration, technology-transfer routes and public-private interaction.

Plant Biotechnology Transfer to Developing Countries,
edited by D.W. Altman and K.N. Watanabe. © 1995 R.G. Landes Company.

INTRODUCTION

Over the past three decades, experience has been gained regarding implementation, integration and management of biotechnology research. Much of this experience has been gained with temperate crops suitable for use in industrialized-country farming practices. Increasing numbers of applications are addressing research objectives for these crops and systems.[1] Relatively little experience exists with the application of biotechnology for tropical crops or farming practices in developing countries.[2] Hence there are few professionals able to assume managerial or planning responsibilities for biotechnology in the national agricultural research organizations (NAROs) of developing countries. However, the number of research activities and scientists using biotechnology is growing in conjunction with the opportunities for training and collaboration provided by national and international biotechnology programs. Thus while research opportunities increase, much knowledge and experience is lacking regarding effective management and implementation of new technologies meeting the needs of developing countries.

In addition, rapid changes have occurred to the global food system over the past decade, which in turn affect the role for agricultural production in the developing countries. These changes, including downsizing of state institutions, liberalization of markets and declining investments in research, have created tensions within established national research and regulatory agencies that handle marketing, export standards and the transition from research to commercial use.[3] This changing environment for agriculture challenges national research leaders as to where to best place emphasis on biotechnology and how to ensure that eventual products fit into the global and national agricultural system.

Specific challenges emerge as research directors and scientists explore approaches for applying biotechnology to agricultural needs of tropical countries and that place them in advantageous positions for global markets. With respect to biotechnology-research program management and implementation, three challenges are described and addressed in this chapter:

1. research collaboration which meets identified needs and objectives;
2. relevant technology transfer routes; and
3. effective public-private interaction.

IBS INTRODUCTION AND APPROACH

The Intermediary Biotechnology Service (IBS) is a special project, based at the International Service for National Agricultural Research (ISNAR), which began in February 1993. It is supported by an international group of donor agencies, including the Directorate General for International Cooperation of The Netherlands and the Swiss Development Cooperation, to act as an independent advisor to national programs in developing countries on matters of biotechnology research management and policy. With regard to challenges facing developing countries, the IBS works with national agricultural research systems (national agricultural research organizations, universities, private foundations and non-governmental organizations) to address the following priorities:

1. assisting NAROs in biotechnology research program management and policy formulation, by developing management and analytical tools;
2. carrying out country reviews to identify priority problems amenable to solution through biotechnology; and
3. identifying and collecting information on international biotechnology program expertise.

A systematic approach has been developed, in the context of conventional agricultural research, for identifying priorities, gaps in planning, relevant policies and managerial and technology-transfer needs which address demand-oriented applications of biotechnology. It provides advice on research program management so that needs are identified for which biotechnology offers a comparative advantage and considers technology-transfer routes available for these research efforts. This approach strengthens management capabilities for integrating biotechnology with conventional agricultural research systems.

To ensure that advice provided to its primary clients reflects a comprehensive look at potential products, users and developers, IBS seeks to work with a range of stakeholders in agricultural research, including representatives of:

- the informal sector, including farming communities, non-governmental or other farmer-represented organizations;
- the formal research sector, including universities, national agricultural research organizations and the international agricultural research centers (IARCs);
- the private/commercial sector; and
- the policy and planning level in national organizations or ministries.

This type of representation is also important with regard to the three management and planning-related issues presented in this chapter. Ensuring such involvement is particularly relevant for technology transfer as knowledge, technologies, materials and germplasm move back and forth through these sectors. It is the cumulative responsibility of these various organizations to supply the best quality inputs in a manner timely for agricultural production cycles. The dissemination of products from biotechnology will take advantage of and build on these same groups.

FRAMEWORK FOR DECISION MAKING

IBS develops its management and analytical tools, inter-related activities and publications in the context of a decision-making framework. Each of these tools is designed to enhance research-management capabilities as related to national agricultural objectives and needs, policies and financing for biotechnology. This approach strengthens national planning capabilities, provides managers and decision-makers with the tools to focus scientists on priorities and time frames, and identifies national policies that stimulate research development, product delivery and technology transfer. Priorities developed within this framework reflect negotiated decisions based on evaluation of information for the points listed above.

The decision-making framework presented considers the national policy environment and the institutional, financial and program issues involved in setting priorities and determining needs for biotechnology-based research.[4] The phases of decision-making are comparable to those used in planning and designing conventional agricultural programs. This provides for the integration of biotechnology to the conventional agricultural research system. The four phases are as follows:

1. setting policies and identifying priorities which address constraints on productivity for which biotechnology offers a comparative advantage;
2. formulating a national program to address these priorities and policies;
3. implementing and monitoring of the research program; and
4. transferring and delivering technologies to end-users (Table 19.1).

This framework recognizes the need for the introduction of knowledge and innovation to agricultural research, and that this can take the form of people, institutions or new technology.[5] With respect to biotechnology, such introductions benefit from multi-level planning, decision-making and policies which focus these contributions on stakeholders' needs. Once such analysis has occurred, the financial investment required to build this knowledge base can be evaluated in a cost/benefit manner and in comparison with other funding demands.

ACTIVITIES AT THE INTERNATIONAL, REGIONAL AND NATIONAL LEVEL

IBS ensures its demand-orientation, relevance and coordination with other programs through a series of integrated actions implemented at the international, regional and national level. This approach helps IBS keep contributions to decision-making in developing countries consistent with national, regional and international planning while providing a systematic means to build needed capacity in agriculture and related sectors.

At the international level, IBS concentrates on a select number of activities and works with other international biotechnology programs to fulfill its intermediary role. These activities include:

- consultations and international symposia which identify and address issues of national importance;
- peer-reviewed publication series: *Biotechnology Research Management Studies*; and

***Table 19.1. Phases of decision-making as identified by IBS
and providing structure for regional policy seminars***

Phases of decision-making	Objectives	Topics for decision-making dialogue and analysis
Phase I: Priority identification and setting national policies	(a) Identify research priorities that are in agreement with production constraints and local needs (b) Identify relevant policy considerations	• Priorities, constraints and needs • Public perceptions, biosafety, IPR • Integration • Financing • Rate of entry
Phase II: Program formulation	Determine program elements for national biotechnology initiative	• Determination of program elements • Scientific review • Cross-sector planning
Phase III: Program implementation and monitoring	Implement the program as designed and develop monitoring system	• Monitoring and evaluation • Socioeconomic analysis
Phase IV: Technology transfer and delivery	Develop strategies to ensure that products reach farmers, growers and consumers	• Product orientation • Technology-transfer routes

- registry of international expertise in biotechnology, assembled to enhance collaboration, and a database (BioServe) available electronically.

At the regional level, IBS brings together the information and reports developed through consultations at the international level with developing countries' participants in policy and planning workshops. The seminars provide for:

- country comparisons and case studies regarding biotechnology undertaken in a regional context following the four decision-making phases (Table 19.1);
- gaps/constraints identification in planning, policies and programs;
- country-specific analysis of biotechnology collaboration through the BioServe database; and
- IBS in-country activities based on analysis of findings.

Working at the national level is the ultimate level of importance for IBS. The regional policy and planning seminars, among other things, will identify follow-up actions for IBS at the national level. These will be pursued with the stakeholders noted above. In-country activities to enhance research program management include:

- strengthening management and planning capabilities for identified needs among the four decision-making phases;
- compile and publish country profiles or reports on priorities, policies and research management issues; and
- initiate proposals which address national needs that are consistent with agricultural planning objectives and enhance international collaboration.

INTERNATIONAL COLLABORATION

IBS identifies opportunities for collaboration with international biotechnology research programs in relation to national priorities and needs for agricultural research. To better understand emerging needs of developing countries with regards to biotechnology planning, and to encourage familiarization with one another's focus and potential for collaboration with international programs, a meeting was organized by the IBS and held at the International Service for National Agricultural Research (ISNAR) in The Hague, The Netherlands. This meeting, titled International Agricultural Biotechnology Programs: Providing Opportunities for National Participation, took place from 9-11 November 1993. Participants included nine representatives of national agricultural research organizations in developing countries, 27 international and regional biotechnology programs and networks, seven multilateral and bilateral donor organizations, and six resource persons invited for specific presentations. The complete report of the results and recommendations from the meeting are available[6,7] as well as more extensive international biotechnology-program descriptions published in a Directory of Expertise.[8]

From this meeting and subsequent review and analysis, IBS constructed its database (BioServe) of international expertise which is now being used in regional and country analysis of the data summarizing objectives and activities of international biotechnology research programs. The data provides a systematic analysis of the international programs and technology transfer or diffusion routes used in developing countries.[9]

Consultations at the international, regional and national level are used to verify IBS country-specific analysis and subsequent actions. This guidance helps strengthen the management and planning capacities in national programs in accordance with their needs and desired research collaboration. In the following sections, this guidance will be examined with regard to the three issues selected for emphasis in this chapter.

MANAGING COLLABORATION WHICH MEETS NATIONAL NEEDS AND OBJECTIVES

An analysis of the international biotechnology research and advisory programs indicated that there are a number of program elements which may be adjusted in relation to program management, goals and collaboration. An analysis of these various program and managerial choices was initiated by IBS through a survey of international biotechnology programs. This included responses from 28 research and advisory programs, six international networks and five donor-agency initiatives.[8]

The analysis of international programs identified seven major program elements and indications of respective percentage of effort. Elements of primary importance are:

- Research and development;
- Human resource development;
- National program participation and networking;
- Program planning, policy and management;
- Monitoring and evaluation;
- Information and communication; and
- Infrastructure development.

These elements were defined as follows. First, research and development included all costs for the actual research component of the programs, whether crop or livestock. Human resource development accounted for training (short- and long-term), including postdoctoral positions. National program participation denotes funding reserved to facilitate research and exchanges with national programs in developing countries. Monitoring and evaluation funding, while limited, gives an important indication of effort planned for the monitoring of biotechnology research. Program planning included internal management issues and their relation to issues such as biosafety. Information and communication documented expenditures for electronic linkages, newsletters and databases. Finally, infrastructure development included, for example, resources for laboratory and computer equipment.

Each program was asked the percent of total effort assigned to the above components. Data received reveal that research and development costs account for approximately 50% of total program costs (Fig. 19.1). In comparison, some 10% is used for national program participation, i.e. those costs available for supporting participation of scientists from developing countries in the international programs. This figure indicates that while research may be well-financed, adequate support to build developing-country collaboration is far less funded.

Opportunities for collaboration with international biotechnology programs have grown, and developing country scientists can now choose among a range of international initiatives. Management recommendations from developing countries are needed to help focus such collaboration and activities on the needs, priorities or programs identified by developing countries. It should be noted that due to individual project objectives and funding support, access to all desirable programs may not be available. The ability for developing country scientists to collaborate with international biotechnology research programs or networks is a key need for managers to consider. Such collaboration breaks down the isolation which scientists may face in developing countries, provides external support for scientific and career development, increases the impact of research through networks and offers a source of peer review.

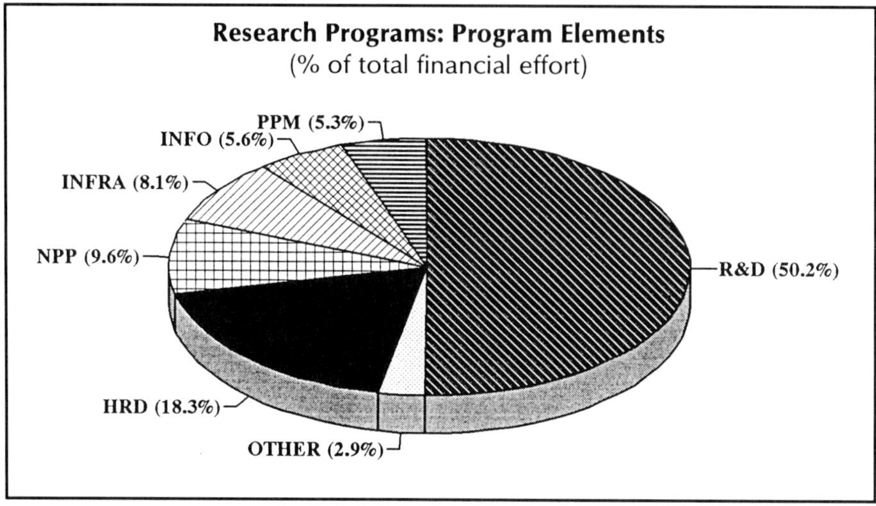

Fig. 19.1. R&D = Research and development; HRD = Human resource development; NPP = National program participation; INFRA = Infrastructure development; INFO = Information and communication; PPM = Program planning, policy and management. Chart is based on data on grant funding for the 25 crop and livestock research programs (totaling US $140 million), in order to keep the information comparable.

Collaborations between South East Asian countries and the international biotechnology programs is shown in Table 19.2. For each program, management responsibilities will include reporting, monitoring and evaluation and the need to factor such research and training into national agricultural research planning. However, if increased collaborative activities are desired, or assistance in related policy matters, then these too must be budgeted and planned in national agricultural development plans. In many cases, planning does not provide for needed funding, as it is hoped or assumed that the international programs themselves will have funds available.

As programs evolve, a shift in relative percent of effort can be expected. However, in order to build expertise in the management of these program and system responsibilities, base-line data on funding strategies is needed. For example, as research activities increase in number and complexity, parallel needs will emerge for training, infrastructure and overall management to assure accountability for funding and activities. As basic needs are met, more funding should become available from the national programs for innovation systems which address technology transfer and product development, as successful research efforts do not automatically lead to improvements reaching the farmer's field.[7]

TECHNOLOGY-TRANSFER ROUTES

Decisions about the production and delivery of products to identified users should be considered early rather than late in structuring

Table 19.2. Collaboration with international biotechnology research activities undertaken in selected South East Asian Countries

Country	# of International Biotechnology Programs	# of Separate Research Activities
Indonesia	12	22
Malaysia	4	9
Philippines	5	16
Vietnam	4	5
Thailand	9	15

(Source: IBS, 1994)

the research program. For national biotechnology initiatives, program formulation and execution are primarily the responsibility of the public sector, with conventional agricultural research remaining the primary conduit for delivering the products of agricultural biotechnology.

Identifying a relevant technology-transfer route is but one factor among many which together help research results reach end-users. Depending on the clients, agroservices and the technology-transfer routes available, the commercial sector may play an important role in production and delivery. Regardless of whether distribution is done by the public or private sector, a route for distribution must be considered as this begins the processes necessary to keep research results from going undeveloped. While national research scientists themselves are not responsible for technology transfer, managers of such research, especially once the research is considered as part of a national biotechnology program, share responsibility for results reaching identified end-users.

There are multiple methods for the transfer of technology into and within developing-country agricultural research institutes. As identified in the IBS analysis of international biotechnology programs, technology-transfer opportunities for biotechnology include the following:
- public sector (NAROs and other government institutions);
- IARCs which release material through international testing programs;
- nonprofit institutions such as universities; and
- commercial organizations.

As shown in Fig. 19.2, the largest effort is expected to occur through the public sector, as most applications of biotechnology supported through the international donor community target crops or production systems traditionally serviced by national extension and research programs. Crops such as rice, beans, potato, sweet potato and cassava, for example, are often planted from seed or planting material saved by farmers. Incentives for large-scale private-sector investment are therefore lacking. For these crops, improved planting materials distributed by IARCs, universities or other international programs will have to be registered for release by each developing country, and this is primarily a public-sector responsibility.

The private sector is playing an increasing role in the transfer of technology into countries through affiliated or licensed research centers, as shown by applications pending for biosafety review in developing countries.[10] However, this technology does not reach the public sector in terms of further research. It is retained by its proprietary developers.

In this regard, national and public institutions also benefit from collaboration with international biotechnology programs. The international programs provide access to both public and proprietary-domain technologies. Examples of commercial technology transfer originating from these international programs is shown in Table 19.3. These new opportunities build on the traditional collaboration of IARCs and developing-country NAROs with public-sector institutions in developed countries for advances in basic research.

These changes reflect the fact that the traditional route of donor, IARC, and developing-country access to biotechnology through public institutions is being affected by the increasing trend towards privatization.[11] This fact, coupled with pressures on national agricultural programs to divest production and distribution responsibilities, may reduce

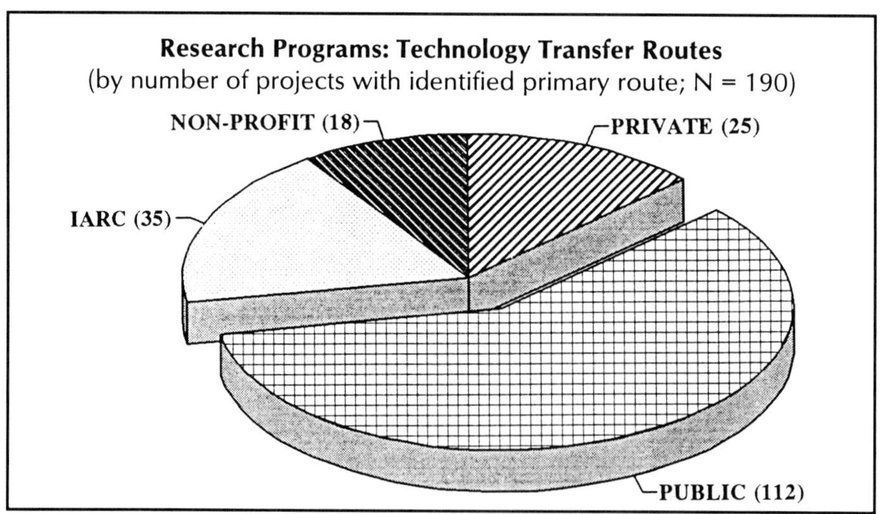

Fig 19.2. PUBLIC = Public sector; PRIVATE = Private sector; IARC = International Agricultural Research Center; NON-PROFIT = Non-profit sector (university, nongovernmental organization). Chart is based on data for 25 international crops and livestock research programs, in order to keep the information comparable.

***Table 19.3. Examples of private-sector technology transfer
in international research programs on agricultural biotechnology***

International Program	Private-sector Collaborator	Technology	Collaborating Institute(s)
Agricultural Biotechnology for Sustainable Productivity (ABSP)	ICI Seeds (USA)	Maize transformation with *Bacillus thuringiensis* protein genes, for resistance to Asian stemborer	Central Research Institute for Food Crops (CRIFC, Indonesia)
	DNA Plant Technology (USA)	Bioreactor technology for micropropagation of banana, pineapple, coffee and ornamental palms	• Agribiotecnologia de Costa Rica (ACR) • Fitotek Unggul (Indonesia)
Feathery Mottle Virus Resistant Sweet Potato for African Farmers (USAID)	Monsanto (USA)	Transformation technology for the development of virus-resistant sweet potato	Kenya Agricultural Research Institute (KARI)
International Service for the Acquisition of Agri-biotech Applications (ISAAA)	Monsanto (USA)	Transformation technology for the development of potatoes resistant to potato virus X and Y	Center for Advanced Research Studies (CINVESTAV, Mexico)
	Asgrow Seed (USA)	Coat-protein technology for the development of melons resistant to cucumber mosaic virus	• Research Center in Cell and Molecular Biology (CIBCM, Costa Rica) • CINVESTAV
	Pioneer Hi-Bred (USA)	ELISA kits for local maize viruses	National Research Center for Maize and Sorghum (CNPMS, Brazil)
ODA Plant Sciences Research Programme	Agricultural Genetics Company (UK)	Insect-resistance genes for potato and sweet potato	International Potato Center (CIP, Peru)

(Source: IBS, 1994)

opportunities for public-sector technology transfer. Commercial producers may be encouraged to assume some of these responsibilities.

If technology transfer through the private sector is an option for the NAROs, then communication with the private sector should occur at an early stage. This helps to ensure that products are appropriate for private production and will be geared to the identified clients or users of the research as identified by the NAROs. In such cases, programs may consider contractual mechanisms for technology transfer, such as collaborative research and development agreements, which itemize the terms of development between public research institutions and private producers.

Consideration of technology transfer and end-user needs becomes more important with increasing research activities. Collaboration between national scientists and one or more international biotechnology research programs accelerates research by providing valuable experience and training needed to build new research capabilities. These capabilities may later be directed towards problems having more specific national importance. Increasing the reciprocal transfer of knowledge and materials between international and national programs and diffusion of products from national programs to end-users requires planning and management of such programs so that technology transfer and diffusion are undertaken as rapidly and safely as possible.

PUBLIC-PRIVATE INTERACTIONS

While many developments in agricultural biotechnology occur in the private or commercial sector, there are relatively few opportunities for effective dialogue between the public and private sectors in developing countries. One of the recognized needs by researchers in developing countries is greater assistance and support for the "development" steps which follow success in the area of research.[6] As highlighted during the meeting, International Agricultural Biotechnology Programs: Providing Opportunities for National Participation, there was a recognized need to increase involvement of the private sector in biotechnology R&D. Regional working groups felt that greater private-sector involvement could be an important catalyst for research funding, product delivery and human resource development. An important weakness identified is in the area of direct linkages between the scientific community and the productive sector. The international programs may need to focus specifically on strengthening public-private linkages in their projects targeted at the regions.

Discussion also focused on the need for product development following initial investments in applied research. Willing and able partners with expertise in product development are difficult to find for developing-country research organizations. While the IARCs and many of the international biotechnology programs described provide support and collaboration for early stages of research, they often do not participate in product development. Lack of support for product development means that many innovations may lack the follow-through needed to move biotechnology from lab to practice. While such issues have been discussed at international meetings,[12] there are relatively fewer opportunities to discuss mechanisms for public-private interaction at the national level.

In order to facilitate greater dialogue and understanding of these matters, IBS includes in its regional policy and planning seminars representatives of both public and commercial agricultural concerns. The composition of the delegations for the IBS policy and planning seminars begins the integration of private sector/commercial expertise and capabilities into the decision-making process for the national research programs. The seminar familiarizes decision-makers with the comparative strengths and advantages of each sector, as well as their comparative abilities in meeting particular agricultural needs. Thus, research directors and decision-makers in the national programs meet their counterparts in the private sector so that research policies and planning can take into account their respective contributions.

Building a broader understanding of how the private sector operates will also increase national research scientists' understanding of intellectual property rights, licenses and material transfer agreements. Following the regional seminars, special attention to these issues and assistance to national programs will be provided if so identified in the regional seminars.

Relatively few of the international biotechnology research and advisory programs were designed to foster collaboration between the public and private sectors. Four of the international programs have been reviewed which provide either conduits for private sector involvement or for their collaboration in research.[7] The involvement of the commercial private-sector in the international dimension of the development-oriented programs is thus relatively small. However, it has been possible to attract matching funds from the private sector which augment those provided by bilateral donors or national programs and universities.[13]

SUMMARY

IBS has initiated a number of follow-up actions building on and incorporating information derived from its experiences with the information presented above. Regional policy seminars are being organized in Asia, Africa and Latin America, which address needs identified as particularly relevant. Special attention is being given to identifying needs of research directors in the NAROs regarding international research collaboration, and strengthening the relations between national and international biotechnology programs. To help in this regard, IBS is building a corresponding database covering information on national efforts in agricultural biotechnology, allowing matching between national needs and international opportunities.

By combining and integrating its various products and seminars, IBS will help build understanding among managers responsible for planning and financing biotechnology in the NAROs. The regional seminars will provide national program leaders with the opportunity for intermediary and advisory activities at the national level, which benefit from close collaboration between IBS, national institutions and international programs. These activities include international collaboration, technology transfer and public-private interactions. Through these activities, national programs become better partners for international collaboration and research by becoming more effective in their management and planning for biotechnology.

REFERENCES

1. Beck CI, Ulrich T. Biotechnology in the food industry: an invisible revolution is taking place. Bio/Technology 1993; 11:895-902.
2. Altman DW. Plant biotechnology transfer to developing countries. Current Opinion in Biotechnology 1993; 4:177-179.
3. Gaull GE, Goldberg RA, eds. The Emerging Global Food System: Public and Private Sector Issues. New York: John Wiley and Sons, 1994.
4. Cohen JI. Biotechnology Priorities, Planning, and Policies: A Framework for Decision Making. A Biotechnology Research Management Study. ISNAR Research Report No. 6. The Hague: International Service for National Agricultural Research, 1994.
5. Crosson P, Anderson JR. Demand and supply: trends in global agriculture. Food Policy 1994; 19:105-119.
6. Cohen JI, Komen J. International Agricultural Biotechnology Programs: Providing Opportunities for National Participation. AgBiotech News and Information, 1994; 6(11):257N-267N.
7. Brenner C, Komen J. International Initiatives in Biotechnology for Developing Country Agriculture: Promises and Problems. OECD Development Centre Technical Paper No. 100. Paris: OECD Development Centre, 1994.
8. IBS. International Initiatives in Agricultural Biotechnology: A Directory of Expertise. The Hague: Intermediary Biotechnology Service, 1994.
9. AgBiotech News and Information. IBS BioServe database shows status of international agricultural biotechnology programmes. AgBiotech News and Information, 1994; 6:19-20N.
10. Napompeth B. (1) Personal communication. (2) Biosafety Regulations in Thailand. Paper presented at the ISAAA Biosafety Workshop, Cisarua, Bogor, Indonesia. April 19-23, 1993.
11. Cohen JI, Chambers JA. Industry and public sector cooperation in biotechnology for crop improvement. In: Moss JP, ed. Biotechnology and Crop Improvement in Asia. Hiderabad: ICRISAT, 1993:17-29.
12. USAID. Strengthening Collaboration in Biotechnology: International Agricultural Research and the Private Sector. In: Cohen JI, ed. Proceedings of a Conference USAID. Washington, D. C: USAID, 1989.
13. Cohen JI. An International Initiative in Biotechnology: Priorities, Values and Implementation of an A.I.D. Project. Crop Science 1993; 33:913-919.

CHAPTER 20

FACILITATING PLANT BIOTECHNOLOGY TRANSFER TO DEVELOPING COUNTRIES

K.V. Raman

INTRODUCTION

In a recent study, the Food and Agriculture Organization (FAO) highlighted the challenges and opportunities provided by biotechnology for developing countries.[1] This study indicates that because of economic considerations, the bulk of biotechnology research is now funded, carried out and controlled by the private sector in the developed world. The biotechnology industry in the U.S. alone invested U.S. $5 billion in research and development.[2] Research institutions in the public sector are now generally required to raise a substantial part of their budget from nongovernmental sources, for example via contractual research, licensing agreements and royalties. This is tending to increase secrecy over research findings and to hinder free scientific communications. University professors, researchers and government institutional scientists are becoming increasingly entrepreneurial and are entering private industry.

Another important trend is that large multinational corporations are purchasing smaller seed and biotechnology companies and diversifying their holdings. The heavy involvement of the private sector as well as market considerations greatly influence the topics and commodities chosen for research. Major crops, commodities and farming systems of great socio-economic importance to the developing world, but of little international market importance, do not figure in the biotechnology agenda of industrial countries. Further, these countries are

Plant Biotechnology Transfer to Developing Countries,
edited by D.W. Altman and K.N. Watanabe. © 1995 R.G. Landes Company.

keen to reduce their production costs and increase the productivity, quality and value of their products, thus improving their overall competitiveness in the world market.

On the other hand, despite the fact that biotechnology facilities are being established in most developing countries, the level of research, development and use of biotechnology for agriculture is generally far below the level in the developed countries. Among developing countries, the status of biotechnology varies considerably. A few countries such as Brazil, China, India, Mexico and the Republic of Korea have good technological and scientific capability in agricultural biotechnology. Many developing countries now face problems of poorly focused research: a shortage of highly qualified personnel, limited access to information, a lack of appropriate policies and priorities, insufficient funding for operational activities, and inadequate linkages between research, development and extension activities. Further, there is a negligible involvement of the private sector which accentuates the problem of insufficient attention to biotechnology.

There are now a handful of specific public programs which are operational to some degree, for transfer of plant biotechnology to developing countries. Activities of these programs are reviewed by Altman.[3] One of these organizations is the International Service for the Acquisition of Agri-Biotech Applications (ISAAA). ISAAA is a non-profit international organization co-sponsored by public and private sector institutions with the aim of facilitating the acquisition and transfer of agricultural biotechnology applications from the industrial countries, particularly proprietary technology from the private sector, for the benefit of the developing world. As a new entity with a specific mission to address the problem of transferring plant biotechnology applications that involve private and public partnership, ISAAA has initiated a program to contribute to the improvement of the current situation in which effective private-public collaborations are infrequent. ISAAA's activities include: (a) assisting developing countries to identify biotechnology needs and priorities; (b) monitoring and evaluating availability of appropriate proprietary applications in industrialized countries; (c) providing "honest broker" services matching needs and appropriate proprietary technologies; (d) mobilizing funds from donor agencies to implement proposals; and (e) counseling developing countries on the safe and responsible testing of recombinant products and assisting in implementing regulatory procedures, planning commercialization activities and assessing impact of the technology. For additional details on ISAAA's objectives and strategy refer to recently published articles.[4-6] Information on project activities of ISAAA were not adequately covered in these reviews. The purpose of this chapter is to provide the current status of ISAAA projects and highlight factors which are critical to the success of transferring agricultural biotechnology applications to developing countries.

PROJECT ACCOMPLISHMENTS

By the end of May 1994, ISAAA had six collaborative projects at the implementation stage. Projects are formulated in three target areas—integrated pest management with transgenic crops, development of diagnostic tools to identify plant diseases, and application of tissue culture techniques for rapid plant propagation. They involve partnerships between universities, commercial firms, and research organizations in Brazil, Costa Rica, Mexico, Thailand, U.S., Venezuela, Zimbabwe, and a network of Asian countries.

Training is an important priority. Under ISAAA's collaborative projects, nine developing country scientists have already received training in biotechnology. Another priority is to help developing nations ensure high standards of biosafety as genetically modified plants are tested and introduced in their countries. All projects negotiated by ISAAA must demonstrate responsible regulatory supervision to protect the environment and to assess risk. In pursuit of this objective, ISAAA has conducted biosafety workshops and collaborated with other international initiatives. These activities have involved several hundred people.

INTEGRATED PEST MANAGEMENT WITH TRANSGENIC CROPS

According to Chasseray and Duesing,[7] 1992 field tests have now been done with 25 transgenic plant and crop species in 21 different countries. These included 393 field trials of transgenic plants completed or planted through the end of 1991. Of the 21 countries, only four developing countries, Argentina, Chile, China and Costa Rica, had tested transgenic plants under their local conditions. Chasseray and Duesing therefore concluded that only about 2% of transgenic field trials occurred in developing countries through the end of 1991. Of this small number, there were virtually none that represented technology transfer which included formal agreements, joint development by donor and recipient partners, and other elements of traditional technology transfer.

Many developing countries now lack biosafety guidelines for evaluating transgenic plants. According to Lesser and Maloney,[8] developing countries are facing the need to make some important and rapid decisions on the safety of biotechnology products. At issue are assurances that experimentally released, genetically engineered organisms represent a minimal threat to the environment and to human health in both the short- and long-term.

Lack of biosafety guidelines in the participating developing countries has therefore been a major limiting factor to technology transfer. ISAAA's current projects in this area therefore focus on helping developing countries to develop appropriate procedures for regulating the testing of transgenic material and to acquire and develop technologies which will serve to demonstrate the comparative advantage and impact of biotechnology on Third World agriculture. The projects in this area include the development of virus-resistant potatoes in Mexico; virus-resistant melons in Costa Rica; virus-resistant papayas in Brazil, Thailand and Venezuela; and insect resistant cotton in Brazil and Zimbabwe. The technologies, nonconventional virus resistance by coat-protein mediated protection and improved resistance by inserting a gene from the bacterium *Bacillus thuringiensis* (Bt) are now well studied and close to the commercialization stage in the developed world. There are numerous benefits to the use of these technologies. Raman and Altman[9] provide additional information on the current status of these technologies and its impact in Third World agriculture.

VIRUS-RESISTANT POTATOES FOR MEXICAN FARMERS

On a global basis, disease losses for potatoes are estimated at 22%—the highest of any major crop. Viral diseases account for much of these losses, particularly in Mexico and other developing countries that lack adequate infrastructure, facilities, and financial resources to operate effective certified-seed programs. No control measures are available for some of these diseases, while for others control depends on repeated insecticide spraying to kill the aphids and other insects that carry disease. Yet insecticides are expensive, often unavailable, and not always effective. Today, Mexican potato farmers have to spray their crops 10 to 16 times during each growing season.

This project was the first to be brokered by ISAAA and was initiated in January 1991. Novel coat-protein technology was donated freely by Monsanto Company to Mexico to confer resistance to two serious diseases caused by potato virus X (PVX) and potato virus Y (PVY). Despite the complexity of the technology Mexican scientists working in conjunction with colleagues in Monsanto Company were able to generate the first transgenic potato variety "Alpha" well ahead of the project schedule, and the material has now been field tested in Mexico. Thus, with the exception of the People's Republic of China, Mexico is the first country in the Third World to have developed and field tested a transgenic crop species. The remarkable progress to date demonstrates that able scientists from developing countries can successfully manage complex biotechnology applications, deliver field products in a time frame hitherto thought to be unrealistic, and are capable of impacting on sustainability in the near-term through increasing food productivity and contributing to a safer environment through reducing dependency on toxic pesticides. The project, funded by the Rockefeller Foundation has now been expanded to include additional varieties with traits to meet more specific needs of poorer farmer groups in the highlands of Mexico, and to address other biotic stresses.

Plans are already underway to formulate, fund and implement the last phase of the project which will be a technical assistance activity to assist Mexico to distribute the transgenic product through cooperatives and other channels. This last phase will be implemented subsequent to release of the transgenic "Alpha" variety by the government of Mexico. For additional details on collaborators and results of this project see the chapter by Rafael Rivera-Bustamante.[10]

VIRUS-RESISTANT MELONS IN COSTA RICA

The local "Criollo" melon is an important food crop for poor farmers in Costa Rica, Mexico and other countries of Central America. Hybrid melons are also an important export crop. However, viral diseases, such as cucumber mosaic virus (CMV), cause significant losses. Today, the only control method is insecticide spraying—as often as 12 times over a 60-day growing season. In 1992 ISAAA launched a project to develop resistance to CMV in "Criollo" melons. Collaborative partners are the University of Costa Rica's Research Center in Cell and Molecular Biology (CIBCM) in San Jose and Asgrow Seed Company, based in Kalamazoo, Michigan (U.S.).

Through this model project, Asgrow donated proprietary technology for producing genetically modified melons with resistance to CMV. Following six-months work at Asgrow Seed Company in the U.S. by a Costa Rican scientist, the procedures for transferring the gene for CMV resistance into "Criollo" melons were accomplished. The second phase of this project focuses on producing and selecting resistant "Criollo" melon lines and testing them in the field. The agreement reached with Asgrow allows Costa Rica to share transgenic criollo melon with Mexico thus facilitating the initiation of South-South cooperation. The project also includes a strong socio-economic component to assess the costs, benefits and broader implications of introducing genetically-conferred virus resistance in an important food crop.

VIRUS RESISTANCE IN PAPAYA: MODEL PROJECT FOR THE TROPICS

Farmers throughout the tropics grow papaya for home consumption and for sale in local markets. Rich in vitamins A and C, papaya is an important dietary component for developing-country farmers and their families. World production is currently estimated at 5 million tons with 98% in developing countries. Throughout the tropics, production is severely affected by papaya ring spot virus (PRSV), which is transmitted by aphids and is extremely difficult to control. No cultivated papaya has ever shown resistance. Working at Cornell University's research center at Geneva, New York (U.S.), Prof. Dennis Gonsalves and his co-workers have developed papayas with good resistance to a PRSV strain found in Hawaii. This application of transgenic technology to tropical fruits is the only example in the world that has advanced to field testing. With technology donated by Cornell, this project is developing papaya lines that resist the PRSV strains found in developing countries of Latin America and Asia.

This project was initiated in August 1992 with the aim of developing bioengineered papaya that are resistant to PRSV for use in Brazil. The first target country is Brazil, the largest papaya producer in the world. A Brazilian scientist from the Brazilian Agricultural Research Corporation (EMBRAPA) identified and isolated a viral-protein gene from the Brazilian strain of PRSV. A second EMBRAPA scientist is now working at Cornell to continue the introduction of the gene into papaya tissues.

EMBRAPA has provided financial support for the development of disease-resistant papaya in Brazil. Funding from other sources will support a similar collaboration with the National Center for Genetic Engineering and Biotechnology (NCGEB) in Thailand and the Universidad de Los Andes in Venezuela. A Venezuelan scientist is currently working at Cornell to produce papaya with resistance to the Venezuelan strain of PRSV. The project will eventually be available for expansion to Colombia, Costa Rica, Honduras, Malaysia, the Philippines, Zaire and other papaya-producing countries. Another crucial component of this project will be a thorough assessment of the impact of introducing a disease-resistant horticultural crop, produced by genetic engineering, on the socio-economic status of subsistence farmers.

PEST-RESISTANT COTTON FOR BRAZIL AND ZIMBABWE

More than two-thirds of the world's cotton acreage is in the Third World, on which over U.S. $1 billion worth of insecticides annually is used which pollutes soils and water resources and places a significant demand on the limited foreign exchange of developing countries. In fact, more pesticides are used on cotton than on any other crop in the world. Brazil is the major cotton producer in Latin America (ranks fourth in the Third World after China, India and Pakistan) and grows about 1.6 million hectares and spends approximately U.S. $150 million/year on insecticides. The high cost of pesticides has driven farmers in many developing countries to abandon cotton production altogether. Bt has been used as an microbial insecticide for more than 30 years, although application is expensive and not entirely effective.

This project involves collaboration between EMBRAPA to develop cotton lines for Brazil with Bt genes already proven to confer resistance to Lepidopteran insect pests. This is ISAAA's first brokered project which will probably result in a joint venture with potential benefits being shared by the two partners, EMBRAPA which has provided the cotton germplasm and Monsanto, the Bt gene. The project is formulated in three stages. The first stage involves the incorporation of the Bt gene through transformation and traditional backcrossing with Bt-engineered germplasm. The second stage will be the field testing of the transgenic material to determine its efficacy and resulting economic benefits. At the completion of the second phase neither Brazil nor

Monsanto has the right to commercialize the jointly-developed transgenic product without the other's consent, and it is envisioned that this will be achieved by ISAAA brokering a joint-venture agreement between EMBRAPA and Monsanto with the financial benefits of commercialization being shared by both parties. A similar agreement to develop transgenic pest-resistant cotton for use in Argentina is now being developed.

In Zimbabwe, cotton is the second most important export crop after tobacco, and the third most important crop in terms of total value after tobacco and maize. In 1993, Zimbabwe produced 226,000 tons of raw cotton. Cotton is particularly important for Zimbabwe's small-scale traditional farmers, the focus of much of the country's rural development efforts—70% of the 1993 crop was produced by more than 100,000 farmers in this sector. Zimbabwe's cotton industry is supported by a substantial breeding program conducted by the Cotton Research Institute which releases new generations of high-yielding varieties every season. The Cotton Marketing Board manages an extremely effective seed-distribution system, supplying even the smallest-scale growers with good planting materials.

The main constraint on productivity are droughts and insect pests. Commercial farmers spend an estimated U.S. $146 per hectare on insecticides, more than 25% of total annual-input costs. When rainfall is normal, cotton growers can expect to spray their crops as many as nine times in one growing season. Subsistence farmers cannot afford this level of expenditure and lose a good deal of their crops to pests. In experiments arranged through ISAAA and conducted at the Cotton Research Institute, Bt proteins provided by Monsanto were tested against the major Lepidopteran insect pests damaging cotton in Zimbabwe. Promising results from these tests are now the basis for negotiations to develop a long-term breeding program for integrating Bt genes into local-adapted cotton varieties. It is possible that Bt can substitute for up to one-third of the annual current costs of cotton insecticides in Brazil and Zimbabwe (up to U.S. $60 million), but the sustainable and environmental benefits associated with less pesticides in soil and water resources will be infinitely more important. Appropriate management of the Bt gene to maximize the durability and reduce the development of resistance is an important element of the project which will ensure a more sustainable deployment of resistance genes.

DEVELOPMENT OF DIAGNOSTIC TOOLS

Diagnostic biotechnologies allow accurate identification of plant diseases, assessment of control strategies and production of disease-free planting material. ISAAA has arranged the donation of a nonradioactive DNA probe to identify black rot, the most important disease of cabbages and other crucifer vegetables in Asian and African countries. Another project involves the donation of monoclonal antibody-identification kits for diagnosing diseases in Brazilian maize crops.

DISEASE DIAGNOSIS FOR CRUCIFER VEGETABLES

Washington State University has developed a nonradioactive DNA probe to detect the bacterium that causes black rot, a disease responsible for major losses in cabbages, broccoli and other crucifers. Black rot is transmitted through seed infected with the bacterium *Xanthomonas campestris* pv. *campestris*. The most effective way to control the disease is to make sure that farmers can buy seed that is disease-free. This requires fast, accurate and inexpensive methods to identify the bacterium in seed before distribution. However, current methods are slow, expensive and not entirely accurate. The new DNA probe can potentially overcome these disadvantages and effectively detect and monitor the movement, extent and seriousness of this important disease, contributing to its control.

Researchers at Washington State University's Research Center have developed a test so sensitive that it can detect as few as one or two individual bacteria in a sample. The Asian Vegetable Research and Development Center (AVRDC) in Taiwan has responsibility for introducing and testing this diagnostic probe in Asia and Africa. Two scientists from this center worked at the laboratory in Puyallup, Washington (U.S.) in 1993 to learn how to make the probe and to develop an adaptive-research phase in Taiwan. Testing and training in the use of the probe is now being implemented at AVRDC.

DIAGNOSING VIRAL DISEASES
IN BRAZILIAN MAIZE CROPS

Maize is one of the most important food crops in Brazil. Annual production is about 21 million tons from 12 million hectares of farm land. This production level is low by international standards, and preliminary assessments indicate losses of up to 11% from viral diseases.

Through an ISAAA project initiated in 1993, Pioneer Hi-Bred International in Johnston, Iowa (U.S.), provided the National Maize and Sorghum Research Center of EMBRAPA in Brazil with virus-detection kits and training based on a well developed biotechnology method called enzyme linked immunosorbent assay (ELISA). ELISA kits produced in the U.S. were tested by Brazilian researchers in training provided by researchers from Pioneer Hi-Bred International for identifying corn stunt spiroplasma and rayado fino virus. As a follow-up, Brazilian scientists now plan to go to Pioneer for extensive training. Financial in-kind support for this project is being provided by EMBRAPA with Pioneer donating the technology and corresponding time required to support the project, including training and visits to Brazil.

APPLICATION OF TISSUE-CULTURE TECHNIQUES

Tissue culture holds particular promise for commercial forestry, reforestation efforts in the tropics, disease control and rapid plant multiplication. By facilitating the mass propagation of well-identified, disease-free plant material, tissue culture can expand and speed up tree-planting efforts, and help to control or reverse environmental and biodiversity degradation. ISAAA now is in the process of negotiating several projects in this area. Promising ones include: bamboo and potato micropropagation in Mexico; accelerated propagation of rattan (*Calamus manan*) in Malaysia; preservation of dipterocarp tree species that are unique in Southeast Asian forests; and micropropagation of pyrethrum and other ornamentals in East Africa.

EMPHASIZING BIOSAFETY

ISAAA has collaborated with several international and national organizations in building institutional capacity in biosafety regulation.[11] With proper guidelines, the genetically-engineered products will be developed, tested and introduced in a safe effective way, and preferably in harmony with existing biosafety regulations in industrial countries. To achieve this objective, ISAAA conducted and collaborated in several biosafety workshops. Workshops conducted in Costa Rica and Argentina in 1992 and in Indonesia in 1993 brought together specialists who will be responsible for oversight in 13 countries of Asia and Latin America: China, Indonesia, Malaysia, the Philippines, Thailand, Argentina, Bolivia, Brazil, Chile, Costa Rica, Mexico, Paraguay and Uruguay. Participants followed real decisions that have approved more than 2,500 transgenic field trials. Additional information on collaborating institutions, and other details on biosafety are given in Raman;[12] Krattiger and Rosemarin.[11] A fourth biosafety workshop is currently being organized for early 1995 to focus on six countries in Africa.

CONCLUSION

ISAAA's limited experience in transfer of biotechnology applications have identified several issues. Technology transfer agreements in this area are complex and involve a number of issues. The issues at stake are mainly the intellectual property protection systems, biological safety and other environmental concerns, substitution of developing-countries' exports, social equity and the widening gap between developed and developing countries in exploitation of the new technologies.

Several aspects have been highlighted by Altman[3] that allowed the success to date. First, there has been an open attitude for developing collaborations, especially involving the private sector, without preconceived ideas or fixed models. In ISAAA's experience, appropriate partners were essential to proceed with the implementation of agricultural biotechnology, and the potential partners in most cases did not have enough familiarity with each other to reach an agreement without an experienced facilitator. Although a facilitator is not a necessity, many of the partners felt that the current situation in biotechnology required this type of assistance. Some specific observations of important components from ISAAA's efforts are: (1) selectivity in choosing limited targets with near-term opportunities being the initial proposal; (2) critical information from knowledgeable individuals on site(s) of application; (3) appropriate technology matching an actual need; (4) immediate attention for regulatory and public-relations issues; (5) a clear entrance and exit policy; and (6) acknowledgment of realistic time lines and cost estimates. As projects get implemented, commercialization activities will likely prove more time-consuming, especially with small farmers. ISAAA will need to facilitate linkages by identifying partners who will assist with protocols for field testing of genetically-engineered materials and help national programs in developing countries make decisions about the appropriateness and extent of commercialization.

ISAAA's philosophy for the near future will continue to promote modern and traditional biotechnologies which will be used as adjuncts and not substitutes for conventional technologies, and their application should be driven by need rather than by technology. In collaboration with other international and national agencies, ISAAA is committed to enhancing the capabilities of the developing countries to enable them to harness biotechnologies in a balanced and equitable fashion, including support to orphan commodities, i.e. those commodities used by the poor. The challenge for ISAAA lies in maximizing the positive effects while minimizing the negative effects of biotechnology.

REFERENCES

1. Anonymous. The State of Food and Agriculture. FAO Agriculture Series 1993. No. 26:64-69.
2. Burrill GS, Lee Jr KB. Accelerating Commericialization, An Industry Annual Report. In: Biotech 1993. San Francisco: Ernst & Young, 1992.
3. Altman DW. Plant Biotechnology Transfer to Developing Countries. Current Opinion Biotech 1993; 4:177-179.
4. Altman DW, James C. Public and Private Sector Partnership through ISAAA for Transfer of Plant Biotechnology Applications. Annals of the New York Academy of Sciences 1993; 700:93-101.
5. Krattiger A, James C. International Organization Established to Transfer Proprietary Biotechnology to Developing Countries: ISAAA. Diversity 1994; 10(1):36-39.
6. Lesser WH, Krattiger AF. Marketing "Genetic Technologies" in South-North and South-South Exchanges: The Proposal Role of a New Facilitating Organization. In: Krattiger AF, McNeely JA, Lesser WH, Miller KR, St. Hill Y, Senanayake R, eds. Widening Perspectives on Biodiversity. Geneva, Switzerland: International Academy of the Environment, 1994:291-307.
7. Chasseray E, Duesing J. Field Trials of Transgenic Plants: an Overview. Agro Food Industry Hi Tech 1992: 3:5-10.
8. Lesser WH, Maloney AP. Biosafety: A Report on Regulatory Approaches for the Deliberate Release of Genetically-Engineered Organisms—Issues and Options for Developing Countries. Ithaca, NY: Cornell International Institute for Food, Agriculture and Development (CIIFAD), 1993.

9. Raman KV, Altman DW. Plant Biotechnology Initiative for International Development to Achieve Pest Resistance. Crop Protection 1994: In press.
10. Rivera-Bustamante R. Example of Transfer of Proprietary Technology from the Private Sector to a Developing Country. In: Altman DW, Watanabe K, eds. Plant Biotechnology Transfer to Developing Countries. Austin, TX: R.G. Landes, 1994.
11. Krattiger A, Rosemarin A, eds. Biosafety for Sustainable Agriculture. Ithaca: ISAAA, 1994.
12. Raman KV. ISAAA's Biosafety Initiative. Proceedings of the USAID Latin American Caribbean Regional Biosafety Workshop. East Lansing, MI USA: Michigan State University: 1993.

AVAILABLE FREE TITLES

Please check three titles in order of preference.
Your request will be filled based on availability. Thank you.

❏ Water Channels
Alan Verkman,
University of California-San Francisco

❏ The Na,K-ATPase:
Structure-Function Relationship
J.-D. Horisberger, University of Lausanne

❏ Intrathymic Development of T Cells
J. Nikolic-Zugic,
Memorial Sloan-Kettering Cancer Center

❏ Cyclic GMP
Thomas Lincoln, University of Alabama

❏ Primordial VRM System and the Evolution
of Vertebrate Immunity
John Stewart, Institut Pasteur-Paris

❏ Thyroid Hormone Regulation
of Gene Expression
Graham R. Williams, University of Birmingham

❏ Mechanisms of Immunological Self Tolerance
Guido Kroemer, CNRS Génétique Moléculaire et
Biologie du Développement-Villejuif

❏ The Costimulatory Pathway
for T Cell Responses
Yang Liu, New York University

❏ Molecular Genetics of Drosophila Oogenesis
Paul F. Lasko, McGill University

❏ Mechanism of Steroid Hormone Regulation
of Gene Transcription
M.-J. Tsai & Bert W. O'Malley, Baylor University

❏ Liver Gene Expression
François Tronche & Moshe Yaniv,
Institut Pasteur-Paris

❏ RNA Polymerase III Transcription
R.J. White, University of Cambridge

❏ src Family of Tyrosine Kinases in Leukocytes
Tomas Mustelin, La Jolla Institute

❏ MHC Antigens and NK Cells
Rafael Solana & Jose Peña,
University of Córdoba

❏ Kinetic Modeling of Gene Expression
James L. Hargrove, University of Georgia

❏ PCR and the Analysis of the T Cell Receptor
Repertoire
Jorge Oksenberg, Michael Panzara & Lawrence
Steinman, Stanford University

❏ Myointimal Hyperplasia
Philip Dobrin, Loyola University

❏ Transgenic Mice as an In Vivo Model
of Self-Reactivity
David Ferrick & Lisa DiMolfetto-Landon,
University of California-Davis and Pamela Ohashi,
Ontario Cancer Institute

❏ Cytogenetics of Bone and Soft Tissue Tumors
Avery A. Sandberg, Genetrix & Julia A. Bridge ,
University of Nebraska

❏ The Th1-Th2 Paradigm and Transplantation
Robin Lowry, Emory University

❏ Phagocyte Production and Function Following
Thermal Injury
Verlyn Peterson & Daniel R. Ambruso,
University of Colorado

❏ Human T Lymphocyte Activation Deficiencies
José Regueiro, Carlos Rodríguez-Gallego
and Antonio Arnaiz-Villena,
Hospital 12 de Octubre-Madrid

❏ Monoclonal Antibody in Detection and
Treatment of Colon Cancer
Edward W. Martin, Jr., Ohio State University

❏ Enteric Physiology of the Transplanted Intestine
Michael Sarr & Nadey S. Hakim, Mayo Clinic

❏ Artificial Chordae in Mitral Valve Surgery
Claudio Zussa, S. Maria dei Battuti Hospital-Treviso

❏ Injury and Tumor Implantation
Satya Murthy & Edward Scanlon,
Northwestern University

❏ Support of the Acutely Failing Liver
A.A. Demetriou, Cedars-Sinai

❏ Reactive Metabolites of Oxygen and Nitrogen
in Biology and Medicine
Matthew Grisham, Louisiana State-Shreveport

❏ Biology of Lung Cancer
Adi Gazdar & Paul Carbone,
Southwestern Medical Center

❏ Quantitative Measurement
of Venous Incompetence
Paul S. van Bemmelen, Southern Illinois University
and John J. Bergan, Scripps Memorial Hospital

❏ Adhesion Molecules in Organ Transplants
Gustav Steinhoff, University of Kiel

❏ Purging in Bone Marrow Transplantation
Subhash C. Gulati,
Memorial Sloan-Kettering Cancer Center

❏ Trauma 2000: Strategies for the New Millennium
David J. Dries & Richard L. Gamelli,
Loyola University

CHAPTER 21

R & D COOPERATION IN BIOTECHNOLOGY WITH DEVELOPING COUNTRIES

Seizo Sumida and Yoshihiko Nishizawa

INTRODUCTION

Japan Bioindustry Association (JBA) is a nonprofit organization dedicated to the promotion of sound development of biotechnology and bioindustry for Japan and the rest of the world. Operated through the support and cooperation of the government, academia and industry, the JBA is a unique organization of well-balanced bases in Japan. JBA's history dates back to 1942 when its forefather, the Japanese Association of Industrial Fermentation (JAIF), was established as a forum for exchange of scientific and technological information on fermentation between academia and industry. In response to the rapid advancement of modern biological sciences and their applications, the organization has evolved, after several re-organizations, into the present form.

Today, like its forefather, JBA functions as a think-tank and forum for communication among scientists, technologists, policy makers and representatives from industry interested in the promotion of sound development and commercialization of biotechnology. The activities of JBA include the following:
1. Promotion of science and technology;
2. Basic feasibility studies for national policy-making;
3. International exchange and cooperation;
4. Socio-industrial harmonization; and
5. Generation, collection and dissemination of information.

Plant Biotechnology Transfer to Developing Countries,
edited by D.W. Altman and K.N. Watanabe. © 1995 R.G. Landes Company.

This chapter focuses on the aspect of international exchange and cooperation, particularly in the context of R&D cooperation with and technology transfer to developing countries.

TRAINING COURSE IN BIOINDUSTRIES

The commitment of JBA to international exchange and cooperation includes its current activities in the program of "Group Training Course in Bioindustries." The activities are an integral part of JBA's technical cooperation and support for developing countries. JBA accepted the requests from the Japan International Cooperation Agency (JICA) in conjunction with the Ministry of International Trade and Industry (MITI) for this program and started its implementation in 1988.

The training course aims to provide experts from research/educational institutions or industry in developing countries with up-to-date knowledge and information on bioindustry. An outline of a typical course follows. The curriculum consists of 18 days of general orientation including Japanese language lessons and 40 days of lectures supplemented with study tours to major scientific and industrial laboratories and factories in Japan. The subjects of lectures include fermentation, plant cell engineering, recombinant DNA technology, bioreactors, cell fusion, purification of biomolecules, biopharmaceuticals, microbial and enzymatic conversion, marine biotechnology, biosensing and bioremediation. Instruction is generally given in English. The participants who successfully complete the course are awarded with a certificate issued by JICA. The government of Japan covers the round-trip airfares and other expenses for accommodations and per diem for all the participants.[1]

By the end of 1993, 47 participants from 13 countries have completed the course. The participating countries have been Argentina, Brazil, Chile, China, Colombia, Cuba, Indonesia, Malaysia, Pakistan, Peru, Philippines, Thailand and Vietnam.

Furthermore, a follow-up study was conducted in Indonesia and Malaysia in 1993 in order to assess the effectiveness of the training program and to identify areas for improvement. The study found that the course program has been welcomed by both countries which shared a common desire for continuation of the program.[2]

R & D COOPERATION
WITH DEVELOPING COUNTRIES

(1) TECHNOLOGY FOR THE CONCENTRATION OF VITAMIN E FROM PALM FATTY ACID DISTILLATE

Background

Palm oil is the second most-abundantly-produced vegetable oil in the world, and more than 60% of the total production took place in

Malaysia in the early 1980s. In light of increasing demand for natural tocopherol (Vitamin E), the development of new natural sources for tocopherol was desirable.

In the research and development cooperation program between the Palm Oil Research Institute of Malaysia (PORIM) and the National Chemical Laboratory for Industry (NCLI) of Japan, scientists found in a laboratory study that natural tocopherol occurred in palm fatty acid distillate (PFAD). As a result, palm oil was considered as a new source of Vitamin E. However, it was found that the chemical composition of palm oil was quite different from that of soybean oil, a conventional source for tocopherol extraction. Therefore, development of refining processes was necessary for the commercial production of Vitamin E from palm oil.

R & D cooperation between Malaysia and Japan

Commissioned by MITI, JBA (JAIF at that time) and PORIM initiated a cooperative project on "Technology for the Concentration of Vitamin E from Palm Fatty Acid Distillate" with a view to accomplishing more effective utilization of palm resources. As part of the Japan-ASEAN cooperation program on science and technology, the project was intended to reinforce scientific and technological cooperation between the two countries.

As mentioned above, this project was based on the outcome of a successful cooperative project on a laboratory scale which was conducted from 1977 to 1980 between PORIM and NCLI of Japan. Elements of further R&D listed below are primarily concerned with scale-up for pilot plant technology:
- Technology for the esterification of PFAD;
- Technology for the concentration and purification of tocopherol;
- Technology for the elimination of the remaining agricultural chemicals from the raw material; and
- Technology for the integration of esterification of raw material and concentration and purification of tocopherols.

The project was implemented according to the schedule (Table 21.1) and successfully completed in early 1988.

(2) INTEGRATED UTILIZATION OF EUCALYPTUS COMPONENTS

Background

Indonesia has one of the largest tropical forest industries in the world with about 60% of its land being forested. The country has been aggressively promoting reforestation programs to cope with vanishing forests and to preserve forest resources. *Eucalyptus* was selected as one of the trees for reforestation because of its rapid growth rate. Consequently, a large number of *Eucalyptus* trees have been planted in Indonesia.

Table 21.1. R&D cooperation on palm oil vitamin E between Malaysia and Japan

Fiscal Year	The schedule of the project implementation
1983	1. General survey of basic program 2. Establishment of the basic plan for the promotion of the Project
1984	1. Preparation of analytical equipment 2. Design work of esterification Pilot Plant 3. Construction of esterification Pilot Plant 4. Design work of concentration Pilot Plant
1985	1. Installation and operation of esterification Pilot Plant 2. Manufacturing, installation and operation of concentration Pilot Plant 3. Design and manufacturing of purification Pilot Plant 4. Integration of Pilot Plants of esterification and concentration
1986	1. Operation of concentration Pilot Plant 2. Manufacturing, installation and operation of purification Pilot Plant 3. Integration of whole Pilot Plant
1987	1. Systematic operation of integrated Pilot Plant 2. Analysis of the data

Because of the quick growth rate, *Eucalyptus* trees were considered as a suitable source for pulp. While the trunks of *Eucalyptus* trees were useful as raw materials for pulp, their branches and leaves were discarded because no use had yet been found for them.

Eucalyptus also was considered as a potentially useful future energy source because of the substantial oil content in its leaves. For this reason, basic research has been conducted on this plant in Japan. It was hoped that, along with the advances of scientific research into chemical components of *Eucalyptus* trees, increasing industrial utilization of their components would be found (Fig. 21.1).

R & D cooperation between Indonesia and Japan

Commissioned by MITI, JBA initiated with the Agency for the Assessment and Application of Technology of Indonesia (BPPT) a cooperative project "Integrated Utilization of *Eucalyptus* Components." Furthermore, as part of the Japan-Indonesia cooperation program on science and technology, it was intended to strengthen scientific, technological and industrial cooperation between the two countries.

The objectives of this project were to carry out research and development of the integrated utilization system and to produce high value-added components from *Eucalyptus* in order to promote effective use of the biomass resources abundant in Indonesia.[3,4] Elements

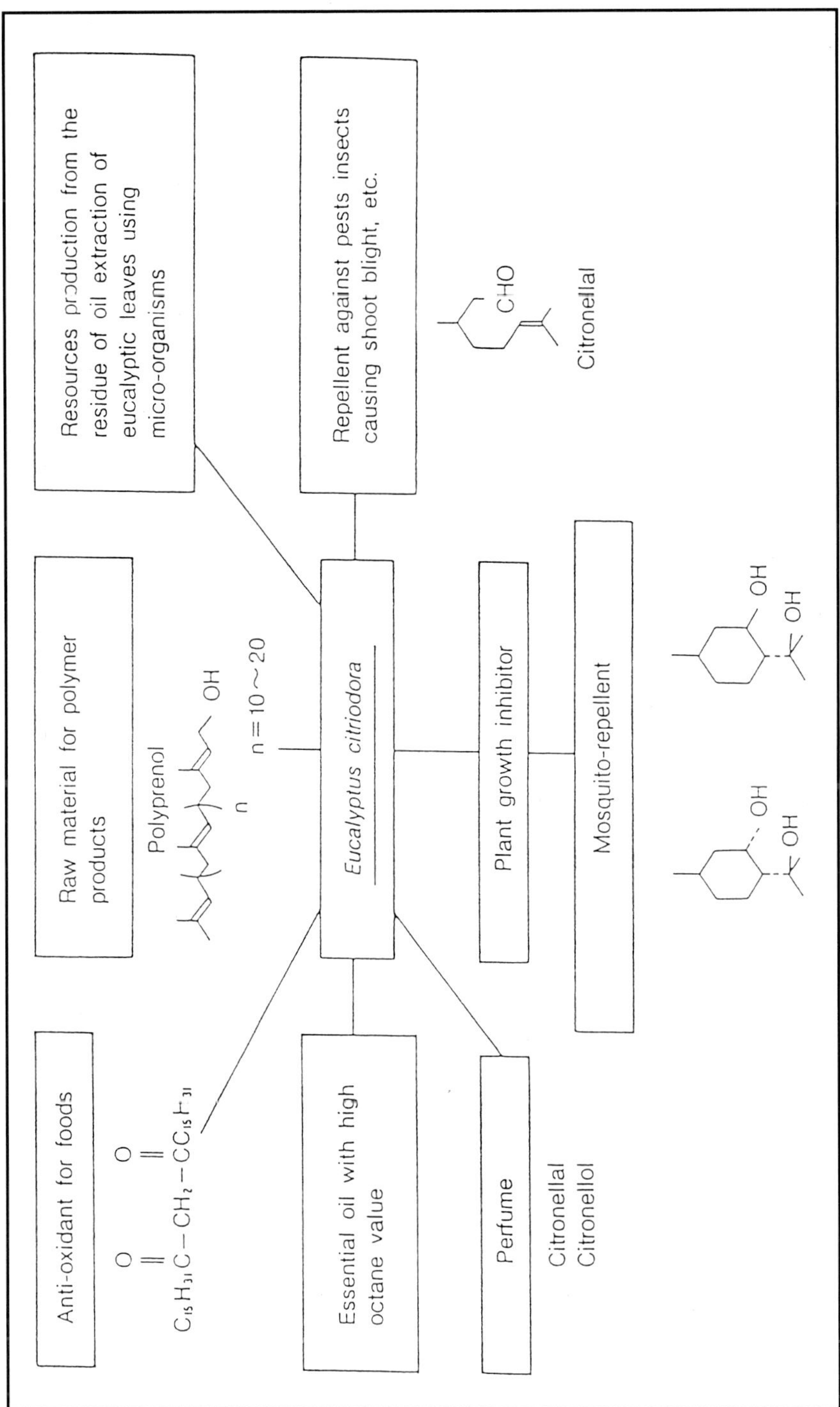

Fig. 21.1. Integrated utilization of Eucalyptus components (an example of Eucalyptus citriodora.)

of the R&D listed below were primarily concerned with scale-up for pilot plant technology:

- Specifying the chemical components of *Eucalyptus* species;
- Pre-treatment processing of leaves;
- Extraction processing of *Eucalyptus* oil and wax from the leaves;
- Separation processing of mixture into oil and wax; and
- Separation and purification processing of useful ingredients from *Eucalyptus* oil.

The project was implemented according to schedule (Table 21.2) and successfully completed in early 1994.

PROMOTING THE CONSERVATION AND UTILIZATION OF BIOLOGICAL DIVERSITY— A PROPOSAL

The high degree of biological diversity in the tropics is evident. For example, the tropical rainforests account for only about 3.3% of the total area of the earth, but about 50% of all organisms are assumed to live there. The conservation of biological diversity is now recognized as a new global challenge. This can be witnessed from the fact that the Convention on Biological Diversity was signed by more than 100 countries, including Japan, at the United Nations Conference on Environment and Development (UNCED) that was held in Rio de Janeiro, Brazil in June 1992.

Table 21.2. R&D cooperation on utilization of Eucalyptus components between Indonesia and Japan

Fiscal Year	Schedule of the project implementation
1989	1. General survey of basic program 2. Establishment of the basic plan for the promotion of the project
1990	1. Preparation of analytical equipment 2. Design work of Eucalyptus Oil, Wax Extraction Pilot Plant 3. Manufacturing of Eucalyptus Oil, Wax Extraction Pilot Plant (a part)
1991	1. Manufacturing, transport and installation of Extraction Pilot Plant 2. Design work of Eucalyptus Oil, Wax Separation Pilot Plant 3. Manufacturing and transport of Separation Pilot Plant
1992	1. Installation and operation of Extraction Pilot Plant 2. Installation of Separation Pilot Plant
1993	1. Operation of Separation Pilot Plant 2. Systematic operation of integrated Pilot Plant 3. Analysis of the data

In order to tackle this issue, the Japan Bioindustry Association organized a committee on "Promoting the Conservation and Utilization of Biological Diversity" under the guidance of the Japanese government. The committee members were composed mainly of experts from academia in diversified fields of specialization. After conducting a study on the conservation and utilization of biological diversity in the tropics, the committee recently prepared a report.[5] The committee hoped that the report will be of help to the Japanese and other governments as well as other organizations in working out plans to contribute to meeting this challenge of world-wide importance. The following are highlights of the report.

In view of what was discussed on the scope and nature of the issue on biological diversity in the tropics, the committee considered it desirable and necessary to establish a "research institute" and "centers" designed for the purpose of conservation of biological diversity in the tropical forests and for the development and transfer of technologies for the effective utilization of biological resources (see Fig. 21.2 for the conceptual framework).

The research institute (referred to by the committee as "International Tropical Bioresources Institute") should function as the core of a global research network. Several regional centers should be built in each major tropical zone and a conservation center in each country in these zones. All these centers in different parts of the world should be linked to form a global research network.

The range of studies to be conducted by the research institute would be exemplified as follows: ecological studies on biodiversity mainly in the tropical forests; technical development for the conservation and multiplication of biological resources; build-up of museum-like functions by exploration and accumulation of biological resources; anthropological studies on ethno-technologies; and biological sciences for the efficient utilization of biological resources. It should publish and exhibit the results obtained through these studies, and transfer the technologies developed to the parties concerned.

The research institute must have a system for international research cooperation and should consult its international advisory board for management and formulation of their basic policy of operation.

The range of departments of the research institute would be exemplified as follows: a research department to be engaged in field and laboratory work involving the exploration of biological resources; a department mainly responsible for the protection and management of the environment, including preservation of bioresources; a museum department for the collection and classification of specimens; a database department for collecting and storing R&D information; an education and training department for publishing study results and conducting technical transfers.

The "regional centers" should formulate and execute research plans as suited to the individual tropical regions.

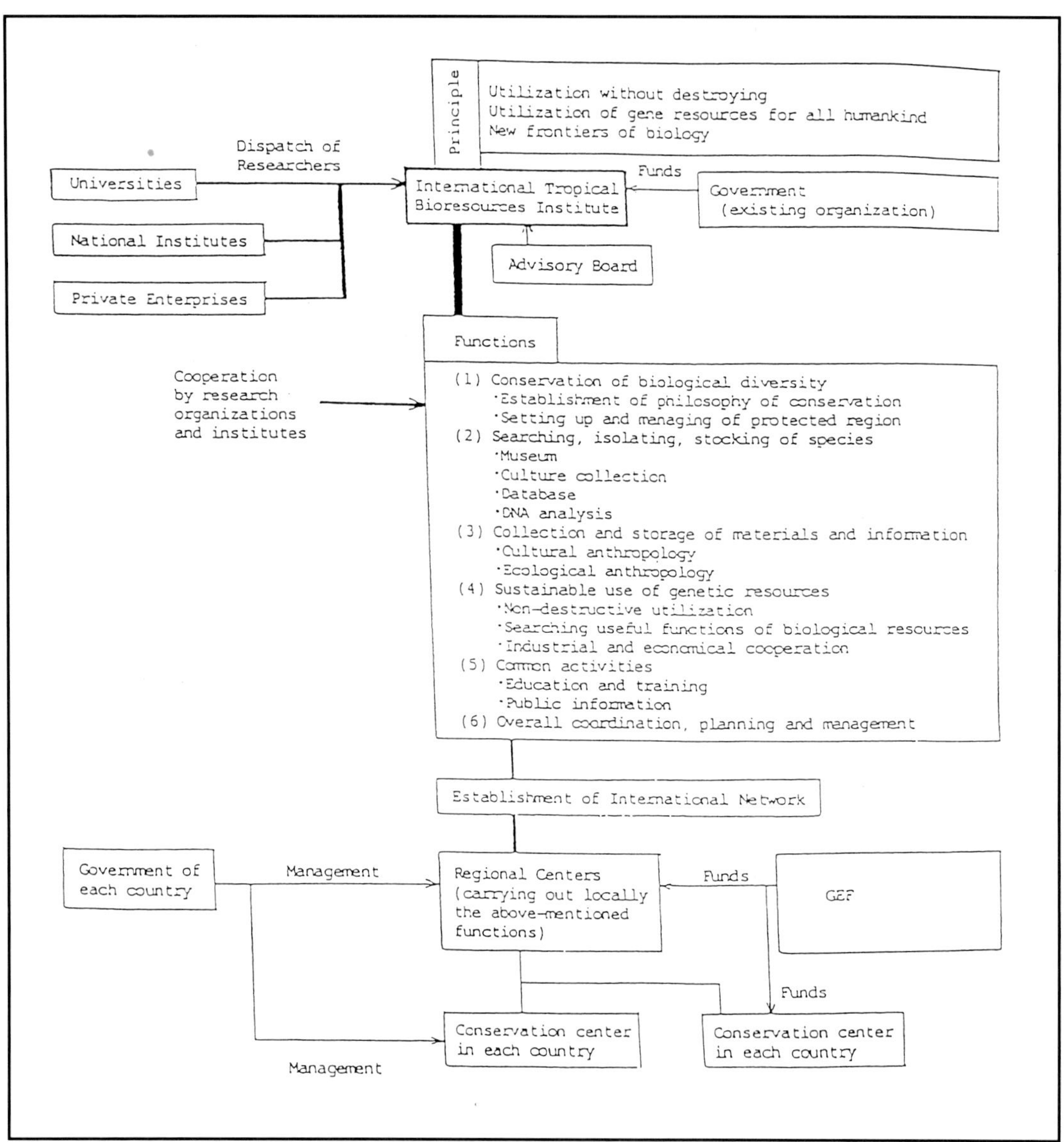

Fig. 21.2. Promoting the conservation and utilization of biological diversity — Conceptual framework.

The proposed concept for the conservation and sustainable utilization of biodiversity has a close bearing on the measures which the biological resource-possessing countries take for their resources. The conceptual framework intends to spark debate on practical initiatives that would have direct implications for plant biotechnology transfer. Further efforts will be necessary to develop this concept into a truly international approach. We believe that there is increasing international awareness about the importance of this cooperative outlook for technology assistance.

REFERENCES

1. The Government of Japan/Japan International Cooperation Agency. Information on group training course in bioindustries.(Course ID: J-94-00357) F.Y. 1994.
2. Sakai S. The developing countries look for group training courses in bioindustries. Japan Bioindustry Letters by JBA 1994; 11(2):8.
3. Nakayama O. Research cooperation in integrated utilization of *Eucalyptus* components. Bioscience & Industry 1993; 51(3):229-30 (in Japanese).
4. United States Patent Serial No. 07/332,238 March 31, 1988.
5. Committee for Promoting the Conservation and Utilization of Biological Diversity in the Tropics. Promoting the conservation and utilization of biological diversity. Digest of Japanese Industry and Technology, 1993; 276:20-30.

Page numbers in italics denote figures (f) or tables (t).

MOLECULAR BIOLOGY
INTELLIGENCE UNIT
AVAILABLE AND UPCOMING TITLES

- Organellar Proton-ATPases
  Nathan Nelson, Roche Institute of Molecular Biology
- Interleukin-10
  Jan DeVries and René de Waal Malefyt, DNAX
- Collagen Gene Regulation in the Myocardium
  M. Eghbali-Webb, Yale University
- DNA and Nucleoprotein Structure In Vivo
  Hanspeter Saluz and Karin Wiebauer, HK Institut-Jena and GenZentrum-Martinsried/Munich
- G Protein-Coupled Receptors
  Tiina Iismaa, Trevor Biden, John Shine, Garvan Institute-Sydney
- Viroceptors, Virokines and Related Immune Modulators Encoded by DNA Viruses
  Grant McFadden, University of Alberta
- Bispecific Antibodies
  Michael W. Fanger, Dartmouth Medical School
- Drosophila Retrotransposons
  Irina Arkhipova, Harvard University and Nataliya V. Lyubomirskaya, Engelhardt Institute of Molecular Biology-Moscow
- The Molecular Clock in Mammals
  Simon Easteal, Chris Collet, David Betty, Australian National University and CSIRO Division of Wildlife and Ecology
- Wound Repair, Regeneration and Artificial Tissues
  David L. Stocum, Indiana University-Purdue University
- Pre-mRNA Processing
  Angus I. Lamond, European Molecular Biology Laboratory
- Intermediate Filament Structure
  David A.D. Parry and Peter M. Steinert, Massey University-New Zealand and National Institutes of Health
- Fetuin
  K.M. Dziegielewska and W.M. Brown, University of Tasmania
- Drosophila Genome Map: A Practical Guide
  Daniel Hartl and Elena R. Lozovskaya, Harvard University
- Mammalian Sex Chromosomes and Sex-Determining Genes
  Jennifer A. Marshall-Graves and Andrew Sinclair, La Trobe University-Melbourne and Royal Children's Hospital-Melbourne
- Regulation of Gene Expression in *E. coli*
  E.C.C. Lin, Harvard University
- Muscarinic Acetylcholine Receptors
  Jürgen Wess, National Institutes of Health
- Regulation of Glucokinase in Liver Metabolism
  Maria Luz Cardenas, CNRS-Laboratoire de Chimie Bactérienne-Marseille
- Transcriptional Regulation of Interferon-γ
  Ganes C. Sen and Richard Ransohoff, Cleveland Clinic
- Fourier Transform Infrared Spectroscopy and Protein Structure
  P.I. Haris and D. Chapman, Royal Free Hospital-London
- Bone Formation and Repair: Cellular and Molecular Basis
  Vicki Rosen and R. Scott Thies, Genetics Institute, Inc.-Cambridge
- Mechanisms of DNA Repair
  Jean-Michel Vos, University of North Carolina
- Short Interspersed Elements: Complex Potential and Impact on the Host Genome
  Richard J. Maraia, National Institutes of Health
- Artificial Intelligence for Predicting Secondary Structure of Proteins
  Xiru Zhang, Thinking Machines Corp-Cambridge
- Growth Hormone, Prolactin and IGF-I as Lymphohemopoietic Cytokines
  Elisabeth Hooghe-Peters and Robert Hooghe, Free University-Brussels
- Human Hematopoiesis in SCID Mice
  Maria-Grazia Roncarolo, Reiko Namikawa and Bruno Péault, DNA Research Institute
- Membrane Proteases in Tissue Remodeling
  Wen-Tien Chen, Georgetown University
- Annexins
  Barbara Seaton, Boston University
- Retrotransposon Gene Therapy
  Clague P. Hodgson, Creighton University
- Polyamine Metabolism
  Robert Casero Jr, Johns Hopkins University
- Phosphatases in Cell Metabolism and Signal Transduction
  Michael W. Crowder and John Vincent, Pennsylvania State University
- Antifreeze Proteins: Properties and Functions
  Boris Rubinsky, University of California-Berkeley
- Intramolecular Chaperones and Protein Folding
  Ujwal Shinde, UMDNJ
- Thrombospondin
  Jack Lawler and Jo Adams, Harvard University
- Structure of Actin and Actin-Binding Proteins
  Andreas Bremer, Duke University
- Glucocorticoid Receptors in Leukemia Cells
  Bahiru Gametchu, Medical College of Wisconsin
- Signal Transduction Mechanisms in Cancer
  Hans Grunicke, University of Innsbruck
- Intracellular Protein Trafficking Defects in Human Disease
  Nelson Yew, Genzyme Corporation
- apoJ/Clusterin
  Judith A.K. Harmony, University of Cincinnati
- Phospholipid Transfer Proteins
  Vytas Bankaitis, University of Alabama
- Localized RNAs
  Howard Lipschitz, California Institute of Technology
- Modular Exchange Principles in Proteins
  Laszlo Patthy, Institute of Enzymology-Budapest
- Molecular Biology of Cardiac Development
  Paul Barton, National Heart and Lung Institute-London
- RANTES, *Alan M. Krensky, Stanford University*
- New Aspects of V(D)J Recombination
  Stacy Ferguson and Craig Thompson, University of Chicago

Neuroscience Intelligence Unit

Available and Upcoming Titles

- ❏ Neurodegenerative Diseases and Mitochondrial Metabolism
  M. Flint Beal, Harvard University

- ❏ Molecular and Cellular Mechanisms of Neostriatum
  Marjorie A. Ariano and D. James Surmeier, Chicago Medical School

- ❏ Ca²⁺ Regulation By Ca²⁺-Binding Proteins in Neurodegenerative Disorders
  Claus W. Heizmann and Katharina Braun, University of Zurich, Federal Institute for Neurobiology, Magdeburg

- ❏ Measuring Movement and Locomotion: From Invertebrates to Humans
  Klaus-Peter Ossenkopp, Martin Kavaliers and Paul Sanberg, University of Western Ontario and University of South Florida

- ❏ Triple Repeats in Inherited Neurologic Disease
  Henry Epstein, University of Texas-Houston

- ❏ Cholecystokinin and Anxiety
  Jacques Bradwejn, McGill University

- ❏ Neurofilament Structure and Function
  Gerry Shaw, University of Florida

- ❏ Molecular and Functional Biology of Neurotropic Factors
  Karoly Nikolics, Genentech

- ❏ Prion-related Encephalopathies: Molecular Mechanisms
  Gianluigi Forloni, Istituto di Ricerche Farmacologiche "Mario Negri"-Milan

- ❏ Neurotoxins and Ion Channels
  Alan Harvey, A.J. Anderson and E.G. Rowan, University of Strathclyde

- ❏ Analysis and Modeling of the Mammalian Cortex
  Malcolm P. Young, University of Oxford

- ❏ Free Radical Metabolism and Brain Dysfunction
  Irène Ceballos-Picot, Hôpital Necker-Paris

- ❏ Molecular Mechanisms of the Action of Benzodiazepines
  Adam Doble and Ian L. Martin, Rhône-Poulenc Rorer and University of Alberta

- ❏ Neurodevelopmental Hypothesis of Schizophrenia
  John L. Waddington and Peter Buckley, Royal College of Surgeons-Ireland

- ❏ Synaptic Plasticity in the Retina
  H.J. Wagner, Mustafa Djamgoz and Reto Weiler, University of Tübingen

- ❏ Non-classical Properties of Acetylcholine
  Margaret Appleyard, Royal Free Hospital-London

- ❏ Molecular Mechanisms of Segmental Patterning in the Vertebrate Nervous System
  David G. Wilkinson, National Institute of Medical Research-UK

- ❏ Molecular Character of Memory in the Prefrontal Cortex
  Fraser Wilson, Yale University